METAMATERIALS

METAMATERIALS

Critique and Alternatives

BEN A. MUNK
Professor of Electrical Engineering, Emeritus
The Ohio State University
Life Fellow IEEE

WILEY

A JOHN WILEY & SONS, INC., PUBLICATION

Published by John Wiley & Sons, Inc., Hoboken, New Jersey.
Published simultaneously in Canada.

For general information on our other products and services or for technical support, please contact our Customer Care Department within the United States at (800) 762-2974, outside the United States at (317) 572-3993 or fax (317) 572-4002.

Wiley also publishes its books in a variety of electronic formats. Some content that appears in print may not be available in electronic formats. For more information about Wiley products, visit our web site at www.wiley.com.

Library of Congress Cataloging-in-Publication Data:

Munk, Ben (Benedikt A.)
  Metamaterials : critique and alternatives / Ben A. Munk.
      p. cm.
  Includes bibliographical references and index.
  ISBN 978-0-470-37704-8 (cloth)
  1. Metamaterials. 2. Antennas (Electronics)–Materials. 3. Electromagnetism. 4. Radio wave propagation–Mathematical models. 5. Antennas (Electronics)–Experiments. 6. Negative refraction. 7. Negative refractive index. 8. Left Handed Materials. 9 Time Advance. I. Title.
  TK7871.6.M855 2008
  621.3028′4—dc22

                                                                    2008030315

Printed in the United States of America

10 9 8 7 6 5 4 3 2 1

This book is dedicated to true science.

The constant support of the ElectroScience Laboratory and my family—in particular my wife, Aase—is deeply appreciated.

A NOTE ON METAMATERIALS

Metamaterials are artificially made materials that do not exist in nature. The term derives from the Greek word *meta*, meaning *beyond*. More specifically, metamaterials are composites that have a desired combination of properties that cannot be obtained by combining the properties of their constituents. The term was coined in 1999 by a colleague and good friend, Rodger Walser, of the University of Texas–Austin, now at Metamaterial, Inc. At my request he graciously provided me with the following definition:

> Metamaterials are macroscopic composites having man-made, three-dimensional, periodic cellular architecture designed to produce an optimized combination, not available in nature, of two or more responses to specific excitation.

I want everybody to understand that I wholeheartedly support a development of metamaterials in general. Only when unrealistic features, in particular a negative index of refraction, are pursued, must I strongly object. Academia, industry, and most urgently, students deserve an honest and frank discussion on this subject. This book has as its focus such a contribution.

B.A.M.

CONTENTS

FOREWORD

Science has always been plagued by occasional hype or misdirected work, witness the N-ray of a previous century that purported to image soft tissues. Unsupported science has appeared in force in recent years in the area of negative parameter materials: for example, "perfect lenses" that cannot produce a usable modulation transfer function and may not even satisfy the standard lens equations. And electrically small antennas enclosed in an NIM (negative index material) shell, "produce a larger voltage"—not a word is said about efficiency, directivity, or bandwidth. These shells have not been constructed or simulated (except in infinite arrays, using a waveguide simulator), and probably never will be, due to mutual interactions inside the shell. Another example is negative refraction through an NIM wedge, where the output beam amplitude shown is normalized to equal that of an equivalent dielectric wedge, even though the very large attenuation (due to the reflection coefficient) might yield new physical understanding. Good science would make measurements of beam shift through a slab, where reflection losses are minimized—but no microwave slab measurements have been made. Much government money is supporting these NIM projects, but the contract monitors are not exercising the careful and skeptical overview that good science requires. Unfortunately, the NIM authors have tried to make *NIM* synonymous with *metamaterials*, the latter term being much broader.

It is important to get a scientific dialogue going; to date the dialogue has been largely one-sided, due to biased journal editorial boards. Professor Munk is initiating a dialogue, with an emphasis on a periodic structure approach, for which he has been a major contributor over many years. Some of the NIM phenomena observed are probably due to surface waves and leaky waves. In his book *Wave Propagation and Group Velocity*, Brillouin shows that signal velocity and group velocity are different for a dispersive medium. Munk reports that the signal velocity always refracts positively, even in an NIM. Almost all of the NIM papers talk about group velocity, but it is really signal velocity that is critical for any practical system. Munk has also been a major contributor to various stealth projects, including absorbers and frequency-selective surfaces. Much of this is explained via Smith charts; he shows that use of these can provide a physical understanding of many complex phenomena.

It is hoped that those responsible for allocating U.S. government research money will find this book useful.

R. C. HANSEN

PREFACE

Why did I write this book? Primarily for two reasons:

1. To show that a lot of statements about certain types of metamaterials, particularly those involving a negative index of refraction, are simply not true
2. To give alternative and, frankly speaking, more realistic solutions to many problems that are widely claimed to be solvable only by using metamaterials (which most often cannot be realized)

The first objective is treated in Chapter 1. Since most attempts to realize Veselago's medium have been made by use of periodic structures, we investigate this subject first. Using relatively simple physical arguments we find that the special features usually associated with Veselago's medium, such as a negative index of refraction and left-handed fields, cannot be realized by a periodic structure. Nor do we find any evidence of waves with phase advance or evanescent waves with increasing amplitude as they move away from their source.

We finally go one step further and take a closer look at Veselago's original paper. It is determined that although his conclusion concerning a negative index of refraction was *mathematically* correct, it suffered from *physical* deficiencies, such as negative time. This is demonstrated by an examination of his flat lens.

However, few people like to read a story with a negative ending (no pun intended). I have therefore gone out of my way to provide examples that supposedly rely on the existence of negative μ and ε but can actually be solved just as well, or better, without them. More specifically, in Chapter 2 we examine a cloak design that is explained through the use of negative permittivity. We show that its performance looked alright but could easily be explained by a simple equivalent circuit based on classical electromotive theory. Further, we suggest an alternative, much simpler design. In this chapter we also look at a case where a short dipole had been tuned to deliver a "higher signal" by surrounding it with a spherical shell made of material with μ, $\varepsilon < 0$. We show that it can be done fundamentally with ordinary materials with μ, $\varepsilon > 0$.

In Chapter 3 we look at structures that absorb at some frequencies whereas they are transparent at others: that is, absorbers with windows, also called rasorbers. Nobody has yet suggested solving this problem by using metamaterials, but I thought it would be nice to beat them to it!

The Russian professor Lagarkov recently considered a structure that could conceptually absorb at almost all angles of incidence. To that end, he considered an absorber placed on top of a slab with $\mu = \varepsilon = -1$. In Chapter 4 we look quite extensively at various absorber designs that actually do quite well what Lagarkov attempts to do, but we use a dielectric slab on top with $\varepsilon \sim 1.6$ and $\mu = 1$. Obviously, our design is more realistic and we show real results up to a $\pm 45°$ angle of incidence.

In Chapter 5 we consider a collinear antenna array that is quite unique in several ways. It has sleeve dipoles placed very close to a heavy mast; in fact, the spacing from the center of the dipoles to the surface of the mast is only $\sim 0.015\lambda$ at the lowest frequency). Such a close spacing has a very strong effect on the terminal impedance of the dipoles, making matching extremely difficult. In this day and age it is often suggested that such a problem be solved by use of magnetic ground planes. Such devices are actually conceptually very simple, but have been elevated to "respectability" by being classified as metamaterials—thus our need to consider such a case in this book. However, magnetic ground planes are inherently very narrowbanded, depending on their thickness; in this case the bandwidth would amount to no more than 1 to 2%. The alternative approach shown here has a bandwidth of 8% not only in one band but in two separate bands an octave apart. More remarkable is the fact that the same matching networks and the same dipole elements were used for both bands. It was, in fact, the most difficult matching problem that I have ever encountered. Thus, it contains so many "neat tricks" that it is unusually suitable as an illustration of classical antenna technology for both students and instructors. But you must know your Smith chart very well to follow this design.

It has been my experience as well as that of several of my colleagues that papers critical of metamaterials are difficult to get published. In fact, when I submitted a paper to the *IEEE Transactions on Antennas and Propagation* in 2003 with the title: "On Negative μ_1 and ε_1: Fact and Fiction," it was turned down vehemently. In it I simply explain how the typical concoction of wires and split-ring resonators acts when treated as a periodic structure without relying on negative μ_1 and ε_1. The intervening years have proven the paper to be correct, and it is given in Appendix A without editing.

Further, in Appendix B we present an interesting IFF antenna mounted in a cavity. It has a steerable cardioid pattern with a front-to-back ratio 15 to 25 dB over a 25% bandwidth. Much antenna design today is done on computers using some sort of metamaterial. This antenna shows an alternative approach: classical antenna design leading to results.

Quite often, transmission lines are needed with a characteristic impedance not readily available (e.g., for matching purposes). Designing a cable of any type is textbook material and consequently, is not discussed here. However, quite often the question is asked: What is a good way to actually measure the characteristic impedance of a cable? Most people think they have the answer, but there are practical pitfalls. Thus, in Appendix C we give a good practical approach that will "alarm" you when there is something suspicious about your measurements.

Quite recently, Hansen wrote a paper that reported on negative refraction apparently observed for a very lossy wedge without negative μ and ε. In Appendix D we consider this problem and show that part of the wave transmitted can, under certain circumstances, be found in the "negative" sector, but it has nothing to do with negative μ and ε.

In closing this preface, I should like to emphasize that I am actually quite tolerant of what I consider a major misconception. I can even tolerate the fact that a great deal of money has been spent on this subject even if it could have been better spent otherwise. However, my deepest concern is that young, talented students are being led into this area without being told that this subject is controversial. In fact, as shown in this book, negative indexes of refraction may not exist. And that implies no time advance and no amplification of evanescent waves.

ACKNOWLEDGMENTS

It is not possible to list everybody who had a direct or indirect part in this book. However, a few stand out and are remembered with gratitude.

Drs. Brian Kent and Stephen Schneider of the U.S. Air Force encouraged me strongly to let my views be known in some written form.

Illuminating discussions as well as suggestions were graciously given by Dr. R. C. Hansen, consulting engineer, Dr. Charlie Rhoads, Raytheon Corp., as well as Dr. Ruth Rotman, Elta Systems, Ltd.

Part of the manuscript was reviewed by Professor Bob Garbacz, Ohio State University and Yeow-Beng Gan, EADS Innovation Works, Singapore. This was followed by numerous discussions that are much appreciated.

Professor Rodger Walser, University of Texas, Austin, opened my eyes to features of metamaterials not seen by me before. Professor Ray Luebbers, Pennsylvania State University, is appreciated for stimulating discussions as well as calculations, as is Professor Leo Kempel, Michigan State University.

Many of my colleagues at the Ohio State University provided valuable input, but two in particular stand out: Professors Jin-Fa Lee and Ron Marhefka, whose help is much appreciated.

Interesting discussions were also had with Professor Jens Munk at the University of Alaska–Anchorage, as well as P. E. Ljung of Stockholm and Dana Quatro.

Finally, the author is indebted to Drs. Peter Munk and Jonothan Pryor, ATK Corporation, for coauthoring Chapter 4. Also, Dr. Clay Larson's (Lockheed) contribution to the IFF antenna in Appendix B is deeply appreciated.

However, no innovations will stay for long unless they are embedded on paper. I am indeed lucky to have secured the help of Lisa Stover, who not only typed the entire manuscript but also made the drawings—and a wonderful job she did. Thank you, Lisa.

B.A.M.

Columbus, Ohio
December 15, 2007

1 Why Periodic Structures Cannot Synthesize Negative Indices of Refraction

1.1 INTRODUCTION

1.1.1 Overview

In this chapter we first list some of the features that are widely accepted as being facts regarding metamaterials with simultaneously negative μ and ε:

1. The index of refraction is negative.
2. The phase of a signal advances as it moves away from the source.
3. The evanescent waves increase as they get farther away from the source.
4. Whereas the E- and H-fields in an ordinary material form a right-handed triplet with the direction of phase propagation, in a material with negative μ and ε, they form a left-handed triplet.

Such materials have never been found in nature. However, numerous researchers have suggested ways to produce them artificially. Periodic structures of elements varying from simple straight wires to very elaborate concoctions have been claimed to produce a negative index of refraction. Nevertheless, we show here that according to a well-known theory based on expansion into inhomogeneous plane waves, it does not seem possible to obtain the characteristic features that are listed above for materials with negative μ and ε. Thus, it seems logical to reexamine Veselago's original paper. We find that it is mathematically correct. However, when used in certain practical applications such as the well-known flat lens, it may lead to negative time. Although such a solution might be acceptable mathematically, it would violate the causality principle from a physical

Metamaterials: Critique and Alternatives, By Ben A. Munk
Copyright © 2009 John Wiley & Sons, Inc.

point of view. So it should not surprise us that, so far, we have encountered difficulties when trying to create materials with negative μ and ε: in particular, a negative index of refraction.

1.1.2   Background

When in 1968 Veselago published his now-famous paper [1], he posed the question: What would happen if a material had both negative permittivity ε and negative permeability μ? Perhaps his most striking conclusion was that a negative sign must be chosen for the index of refraction:

$$n_1 = -\sqrt{\mu\varepsilon} \qquad (1.1)$$

This observation led to significant new concepts. We list the most important in Section 1.2. We emphasize that at this point we neither endorse nor condone these new concepts. However, subsequently, in Sections 1.4 to 1.6, we investigate whether it is feasible to synthesize Veselago's material by the use of periodic structures made with special elements. We will find this to be highly unlikely. In view of this, in Section 1.10 we investigate whether Veselago's conclusion violates fundamental physical principles.

Further, in Section 1.9 we examine the dispersion of a cable terminated in a complex load. We show that in that case it is indeed possible to partially eliminate dispersion over a limited frequency band. This is equivalent to the mixture of forward- and backward-traveling waves deemed essential to achieve the special features of Veselago's medium. However, it is erroneous to conclude that a new exotic material has been created. It will simply lose its features if the load impedance is, for example, purely imaginary. More specifically, we have merely used old tricks from broadband matching techniques.

1.2   CURRENT ASSUMPTIONS REGARDING VESELAGO'S MEDIUM

1.2.1   Negative Index of Refraction

In his original paper, Veselago [1] concluded that the index of refraction n_1 between an ordinary medium and one with negative ε and μ would be negative. Thus, as illustrated in Figure 1.1, the refraction angle θ_r would, according to Snell's law, have the same sign as the angle of incidence θ_i when $n_1 > 0$, whereas it would be negative for $n_1 < 0$. Veselago's original proof is discussed in Section 1.10.

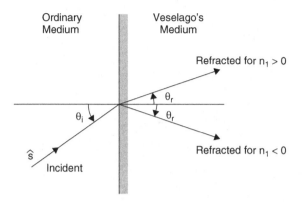

Figure 1.1   Snell's law for an ordinary medium adjacent to Veselago's medium for index of refraction $n_1 > 0$ and $n_1 < 0$, respectively.

1.2.2   Phase Advance when $n_1 < 0$

If a lossless dielectric slab is placed in front of a ground plane, the input impedance Z_i for an ordinary material with $n_1 > 0$ will be obtained by a rotation $2\beta d = 2\beta_0 n_1 d$ in the clockwise direction, as shown in the Smith chart in Figure 1.2. Similarly, if $n_1 < 0$, Z_i is obtained by rotation in a counterclockwise direction. In other words, we experience a phase delay when $n_1 > 0$ and a phase advance when $n_1 < 0$. These statements are based on refs. 2 to 4. Note that loss is not necessary to obtain these features.

1.2.3   Evanescent Waves Grow with Distance for $n_1 < 0$

When propagating waves change into evanescent waves, it is usually because n_1 goes imaginary [5]. Thus, in view of the phase advance postulated above, it should not surprise us that Pendry [6] suggested that evanescent waves in a medium with $n_1 < 0$ would grow and not be attenuated as usual for $n_1 > 0$, as illustrated in Figure 1.3.

1.2.4   The Field and Phase Vectors Form a Left-Handed Triplet for $n_1 < 0$

Also shown by Veselago in his original paper [1] was that the field vectors $\bar{E}$ and $\bar{H}$ and the direction of phase propagation $\hat{s}$ form a left-handed triplet when $n_1 < 0$ (see Figure 1.4b). This feature is probably the least observed when performing experiments. However, as we shall see later, it is a theoretical point very powerful in determining whether or not we have a true Veselago medium.

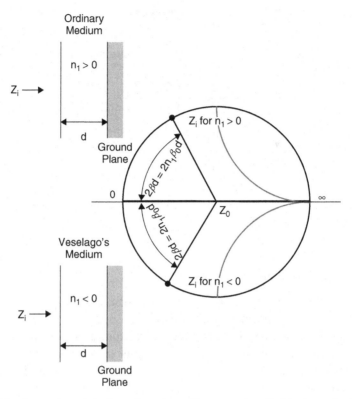

Figure 1.2   Perception of the input impedance Z_i as seen in a Smith chart of a dielectric slab in front of a ground plane for index of refraction $n_1 > 0$ (top) and for $n_1 < 0$ (bottom). For a discussion about causality for $n_1 < 0$, see equations (1.15) to and (1.17).

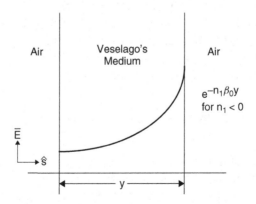

Figure 1.3   Normally, an evanescent wave is attenuated as it moves away from its source. In Veselago's medium it is believed to grow.

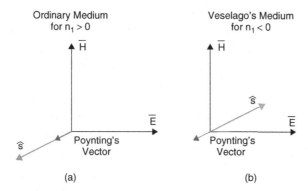

Figure 1.4   (a) In an ordinary medium, $\bar{E}, \bar{H}$, and the direction of propagation $\hat{s}$ form a right-handed triplet; (b) in Veselago's medium, $\bar{E}, \bar{H}$, and the direction of propagation $\hat{s}$ form a left-handed triplet. However, Poynting's vector always points in the same direction.

1.3   FANTASTIC DESIGNS COULD BE REALIZED IF VESELAGO'S MATERIAL EXISTED

When this author started to design high-precision antennas more than 50 years ago, he quickly realized that if the input impedance of a transmission line could go backward in the Smith chart with increasing frequency, matching antennas would, in general, be trivial. He also quickly observed that such components were just not available. However, that was the essence of what the Veselago material promised (if realized). Thus, it is no wonder that an avalanche of papers appeared (mostly simulated), all based on the assumption that Veselago's material was indeed possible to realize.

The most prominent concept was probably the flat lens, discussed in Section 1.10.3. Further, when Pendry later suggested that the evanescent waves at the source would arrive more strongly at the image (see Figure 1.3), the enthusiasm almost boiled over. The possibility of obtaining an optical system that could exceed the traditional diffraction limits was undoubtedly one of the greatest factors that kept funding going for years.

Similarly, Engheta gave a paper in 2001 in Torino [3] in which he considered the resonance frequency of a cavity between two ground planes. He suggested that the space was filled partly with ordinary dielectric with $n_1 > 0$ and the remainder with material with $n_1 < 0$. It was also stated that consultation of Figure 1.2 would readily show that the resonance frequency could potentially remain constant from dc to broad daylight! (The two ground-plane impedances could essentially cancel each other, regardless of frequency.) When the present author pointed out from the floor at the meeting that the rotation in the Smith chart for $n_1 < 0$ violated Foster's reactance theorem, it by no means settled the issue. In fact, it resulted in another showing that for $n_1 < 0$, a modified Foster's reactance theorem would indeed indicate counterclockwise rotation in accordance with Figure 1.2 [4]. The question was, and is, of course: Is there a material with $n_1 < 0$? Veselago himself was quick to point out that his material had never been found in nature. And he added, prudently, that there were perhaps profound reasons for its absence.

1.4   HOW VESELAGO'S MEDIUM IS ENVISIONED TO BE SYNTHESIZED USING PERIODIC STRUCTURES

For almost 30 years after Veselago published his original paper, there was little evidence of any particular interest in his material. However, in the mid-1990s, Pendry postulated that a negative ε could be produced by a periodic structure of strips, as shown in Figure 1.5a. Actually, such a surface is usually found to be inductive [5, Chap. 1]. However, an inductor

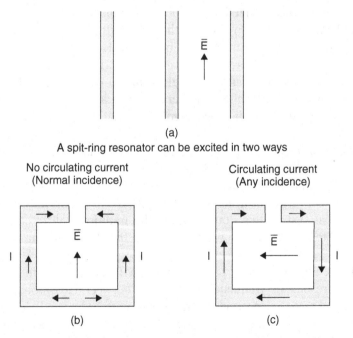

Figure 1.5   (a) Pendry suggested that a negative ε could be produced by an array of parallel wires; (b) and (c) similarly, a negative μ is expected from an array of loops with circulating currents.

can also be considered to be a negative capacitance, which again indicates the presence of a negative ε. Later, Pendry suggested that a negative μ could be obtained from a periodic structure of open split-ring resonators, as shown in Figure 1.5b and c. The idea here was that a circulating current was able to produce a negative μ [6–9]. However, we should note that the current induced is highly dependent on the orientation of the incident E-field. In the case shown in Figure 1.5b, the incident E-field is vertical, which for normal incidence will produce only push–push currents, as indicated in the figure, whereas for oblique incidence in the horizontal plane a weak circulating current will be present in addition to strong push–push currents. However, when the incident E-field is horizontal, as shown in Figure 1.5c, we will observe a circulating current for any angle of incidence unless $\bar{E}$ is perpendicular to the plane of the loop.

It was not long after Pendry's postulates that a group of physicists at the University of San Diego made a combination of flat wires and split-ring resonators, as shown in Figure 1.6 [10–13]. They then performed measurements on a wedge-shaped body as shown in Figure 1.7b. The idea was, as illustrated, that the refracted field would depend strongly on the sign of the refractive index, n_1. In fact, they measured the refracted

Figure 1.6 Original periodic structure used by the San Diego group to demonstrate the presence of negative refraction.

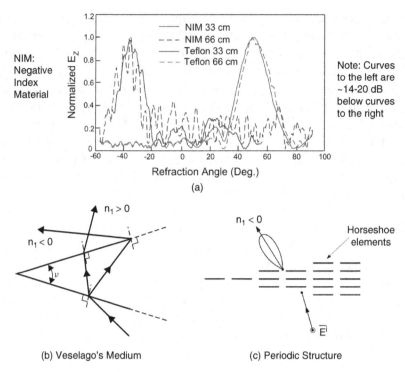

Figure 1.7   (a) Curves to the right represent the field refracted through a Teflon wedge as shown in part (b). Curves to the left are perceived as being the field refracted through a wedge made of wires and a split-ring resonator, as shown in part (c). Note that they are actually about 20 dB below (ca. 1% power) the curves to the right even if they are all shown normalized to 100%. (After ref. 14, with permission.)

field for a Teflon wedge ($\varepsilon \sim 2.1$) and obtained the refraction curve to the right in Figure 1.7a. They also measured the refracted field from a wedge-shaped assembly of wires and split-ring resonators, as shown in Figure 1.7c. Actually, the (measured) curves to the left in Figure 1.7a were not measured by the San Diego group but were obtained later by a group working at Boeing's "Phantom Works" [14]. They went to great lengths to obtain the exact refraction in both the far field [NIM (negative index material) 66 cm] and the near field (NIM 33 cm). Note how the sidelobes in the far-field pattern are almost gone for the near-field case, as is typically seen in antenna experiments. However, the most interesting feature is probably the fact that the refracted field for the synthesized material is about 14 to 20 dB or more below the refracted field for the Teflon case. Such a large loss cannot be attributed to either ohmic or dielectric loss for frequencies below 100 GHz. This fact and the presence of sidelobes in the far field suggested to this author that the refracted field for the synthesized material was actually not a refracted field but merely the radiation pattern for a surface wave that can exist only on a finite periodic structure. Such surface waves have not been demonstrated for split-ring resonators per se. However, they are well documented for simple straight wires (dipoles) [15–21]. In fact, a typical surface wave is shown in Figure 1.8b, where 50 dipoles, each of total length about 0.35λ, are exposed to an incident plane wave at 45°, as shown in the insert [20,21]. First, we note that the level of the surface-wave radiation is about the same as that of the blue refracted curve in Figure 1.8a (ca. 20 dB down). Next we note that the decay rate for the sidelobes is about the same for the two patterns. It should be emphasized further that the orientation of the E-field is as indicated in Figure 1.5b (see Figure 1.6). Thus, there were only very weak circulating currents such that μ would be weak according to Pendry. Nevertheless, negative refraction, although very weak (<14 to 20 dB below a Teflon wedge), was still claimed.

Finally, there are numerous papers in which negative refraction has been claimed for basically straight loaded or unloaded elements with no circulating currents, typical examples being shown in Figure 1.9 [22,23]. Note, in particular, Figure 1.9c, where the elements have been printed on each side of a thin substrate and the elements flipped to avoid any possible chiral or loop effect. All of these elements claim to have measured negative index of refraction, although with more than 20 dB loss.

The discussion above does not constitute a proof of whether we actually observe negative refraction or witness another phenomenon. We have suggested here that it is quite likely the radiation from a surface wave that typically exists over about 10% bandwidth. However, it could also simply be part of the sidelobes from the main beam of the field transmitted. Anyone with experience in measuring the fields scattered from a periodic structure will know how difficult such measurements are: in particular, if we are down 20 dB or more. In the next section we show that this phenomenon is almost certainly not due to refraction.

1.5 HOW DOES A PERIODIC STRUCTURE REFRACT?

1.5.1 Infinite Arrays

In this section some simple and well-known facts about periodic structures are pointed out. Unfortunately, they are too often overlooked, forgotten, or simply ignored! Consider an infinite × infinite array, as shown in

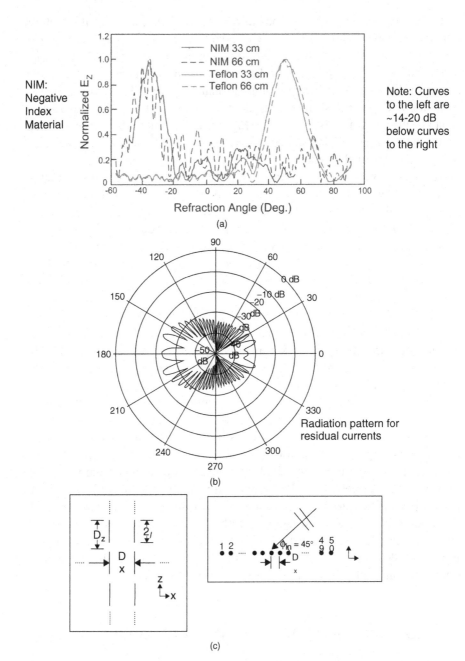

Figure 1.8 (a) Same curves as shown in Figure 1.7. However, note how the sidelobes of the curves to the left are similar to the sidelobes of surface-wave radiation shown in part (b). Further, the radiation intensity is about the same (ca. 20 dB below maximum).

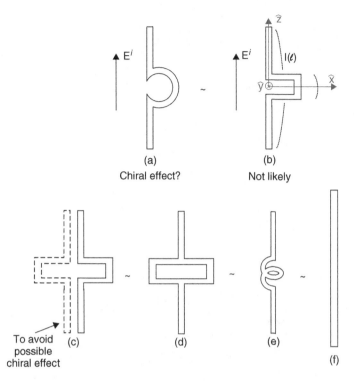

Figure 1.9   Some of many elements with and without circulating current where a negative index of refraction has been claimed, always 20 dB or more below reference.

Figure 1.10. It is being exposed to an incident plane wave with direction of propagation $\hat{s}$. For the sake of simplicity we will for the time being assume $\hat{s}$ to be contained in the yz-plane. For $\hat{s}$ pointing upward to the right as shown in the figure, it is clear by implication of Floquet's theorem that the voltages induced in row 1 will be delayed by $\beta D_z s_z$ compared to row 0. However, the fields re-radiated from row 1 will be ahead by the same amount, $\beta D_z s_z$, for waves propagating in the forward direction $\hat{s}$ as well as in the specular direction $\hat{s}_s = \hat{x} s_x - \hat{y} s_y + \hat{z} s_z$, as illustrated in Figure 1.10a and b, respectively. In other words, propagation in these two directions is always possible unless the element pattern has a null in any of these directions.

We now ask: Is it possible to reradiate a plane wave in an arbitrary direction $\hat{s}_a$? If so, the elements in row 1 will have a phase advance of $\beta D_z s_{az}$. Only if the sum of the delay and advance adds up to a multiple of 2π can a plane wave propagate in the direction $\hat{s}_a$. (Remember: Our array is infinite $\times$ infinite, not finite; see later.) Thus, the condition for

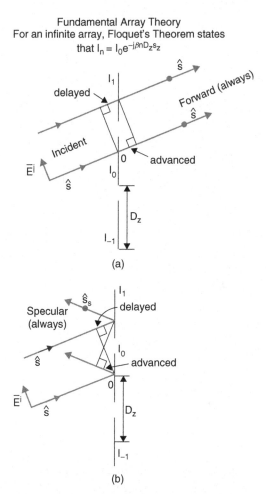

Figure 1.10 An incident plane wave with direction of propagation $\hat{s}$ will induce a voltage in element 1 that is delayed by $\beta D_z s_z$ compared to element 0. Conversely, the re-radiated field from element 1 will be advanced by $\beta D_z s_z$ compared to element 0 in the forward direction $\hat{s}$ (top) as well as in the specular direction $\hat{s}_s$ (bottom). Thus, propagation in the forward and specular directions is always possible. (Note the infinite arrays.)

reradiation in the arbitrary direction $\hat{s}_a$ is (see Figure 1.11)

$$\beta D_x (s_z - s_{az}) = 2\pi n_1 \qquad n_1 = 0, \pm 1, \pm 2, \ldots$$

or recalling that $\beta = 2\pi/\lambda$,

$$\frac{D_z}{\lambda} = \frac{n_1}{s_z - s_{az}} \qquad n_1 = 0, \pm 1, \pm 2, \ldots$$

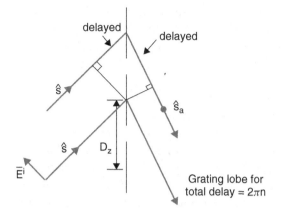

Grating lobes possible only for $D_z > \lambda/2$.
Thus, NO negative refraction for "continuous"
medium (i.e., $D_z < 0.4\lambda$)

Note: Floquet's theorem is valid for ALL
element types (they merely change the
element pattern). This is at variance with the
theory for an artificial dielectric.

Figure 1.11 In contrast to the case in Figure 1.10, propagation in an arbitrary direction $\hat{s}_a$ is possible only if the total phase delay is $2\pi n$. These are simple grating lobe directions. (Note the infinite arrays.)

which shows that it can always be satisfied provided that we make D_z/λ sufficiently large. However, the smallest value of D_z/λ is obtained for $n_1 = +1$ and $s_z = 1$ (grazing incidence upward) and $s_{az} = -1$ (grazing re-radiation downward). In that case,

$$\frac{D_z}{\lambda} = \frac{1}{2}$$

In other words, for $D_z/\lambda < \frac{1}{2}$, re-radiation is possible *only* in the forward direction $\hat{s}$ and the specular direction s_s (infinite array only). For $D_z/\lambda > \frac{1}{2}$, propagation in other directions is possible (see Figure 1.11). In fact, these are simply the well-known grating lobe directions.

Note that the phase velocity along the z-direction is opposite for the incident and the lowest grating lobe direction. For that reason this grating lobe has sometimes mistakenly been denoted as a "backward"-traveling wave. These grating lobes are encountered in numerous microwave devices, such as the backward-traveling oscillator and the backward-traveling antenna, as well as in photonic bandgap materials. They have been known for a long time and are well understood. Again, we emphasize that these backward-traveling waves can exist only when the interelement spacings D_x and D_z exceed $\lambda/2$, and not for D_x, $D_z < \sim 0.4\lambda$.

The term *backward-traveling wave* was later suggested to mean a wave where the phase and group velocity were opposite each other [28–31]. It was thought that Veselago associated such waves with his findings for media with negative μ and ε when he concluded that the phase velocity and the Poynting vector were opposite each other. However, this writer is not aware that he ever used the term *backward-traveling wave*. Note: The grating lobes have identical phase and group velocity in a dispersionless medium. There is nothing "backward" about them.

We should note further that in the world of metamaterials, the interest seems to concentrate on two types of materials:

1. The interelement spacings D_x and D_z are somewhat smaller than $\lambda/2$, typically $\lambda/4$ or smaller. These cases are often denoted as "continuous," and the phase difference between adjacent elements is typically ignored (an approximation not allowed in the rigorous theory of periodic structures).

2. The interelement spacings are somewhat larger than $\lambda/2$, typically 0.7 to 1.5λ. These materials fall into a category usually called *photonic bandgaps* or *crystals*. They are often perceived as being able to propagate "backward"-traveling waves. Actually, these are nothing but grating lobes, as discussed above.

In other words, *the direction of refraction in air is determined solely by the interelement spacings D_x and D_z as well as the direction $\hat{s}$ of the incident field, never by the element type.*

These will determine where the structure resonates, the bandwidth, and to some extent, variation with angle of incidence as well as the amplitude in general of the scattered fields. Which leaves us with the following conclusion: The extensive discussion of whether surfaces with elements such as the split-ring resonator are blessed with negative refraction and others are not is somewhat misguided. In fact, a periodic structure (of infinite extent) of the continuous type $D_x, D_z < \lambda/2$ and no dielectric can only produce a refracted field with refraction at $n = +1$! You may forget entirely about negative refraction!

1.5.2   What About Finite Arrays?

The categorical denial above that any refraction other than the forward is specular if $D_x, D_z < \lambda/2$ is, rigorously speaking, true only for an infinite structure. In reality, all structures are, of course, finite. This fact will have certain consequences. Foremost, the signals in the forward and specular directions will occur in the form of main beams in each of these directions. They will be flanked by numerous sidelobes. Exact calculated examples of finite × infinite arrays is given in refs. 15–21.

Further, a finite periodic structure is able to sustain certain types of surface waves not possible when the same structure is of infinite extent. Note: It is of utmost importance that the interelement spacing be less than $\lambda/2$ (i.e., the structure is of the "continuous" type). In that event we find that currents associated with the surface wave can be much stronger than currents associated with the mainbeams described above, typically over about 10% bandwidth and when the total element length is about 0.35λ for a simple dipole element [20; Figure 10 in Appendix A]. The surface wave current will, of course, re-radiate like any other element current. The good news is that the surface wave has a low radiation efficiency such that the reradiated field is typically about 14 to 20 dB or more below the amplitude of the mainbeam despite the higher current amplitudes. In addition to the pure surface waves, some currents will usually be associated with reflections from the edges. However, these radiations are usually small compared to those of pure surface waves (for details, see ref. 20).

The sum of the surface wave and the end currents are often referred to as *residual currents*. The re-radiation from these is shown by the radiation pattern in the middle of Figure 1.8b. Note that it has both the same level as the blue curves (about 20 dB below the Teflon wedge) and similar sidelobes: in short, a strong indication that we are seeing radiation from a surface wave and not a simple refraction. (In that case there would be *no* sidelobes!)

The discussion above emphasized the physical aspect of refraction. However, for those who prefer a more mathematical approach, in the next section we present the highlight of the plane-wave expansion [5]. This will demonstrate essentially two features:

1. The re-radiated field from a periodic structure is always right-handed, regardless of element shape or type.
2. The field both inside and outside a multilayered periodic medium is always right-handed.

1.6   ON THE FIELD SURROUNDING AN INFINITE PERIODIC STRUCTURE OF ARBITRARY WIRE ELEMENTS LOCATED IN ONE OR MORE ARRAYS

1.6.1   Single Array of Elements with One Segment

Consider a single planar array as shown in Figure 1.12. The elements are oriented along $\hat{p}^{1,1}$, where $\hat{p}^{1,1}$ is arbitrary except that it is contained in the plane of the array.* Further, we denote the infinitesimal element length by $dl^{1,1}$, the current by $\mathbf{I}^{1,1}$, and the reference point of the reference element by $\bar{R}^{1,1}$. This array, with interelement spacings D_x and D_z, is exposed to an incident plane wave with direction of propagation

$$\hat{s} = \hat{x}s_x + \hat{y}s_y + \hat{z}s_z \tag{1.2}$$

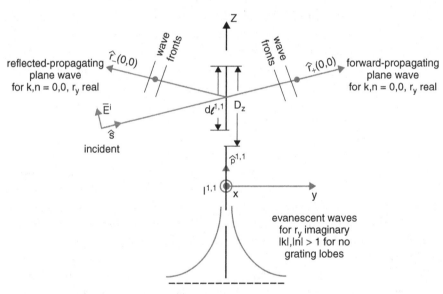

Figure 1.12   Plane wave with direction of propagation $\hat{s}$ incident upon an infinite array of single-segment elements with orientation $\hat{p}^{1,1}$, length $dl^{1,1}$, current $I^{1,1}$, and reference point $\bar{R}^{1,1}$. A plane wave will be scattered in the forward direction, $\hat{r}_+(0,0) = \hat{s}$, as well as the specular direction, $\hat{r}_-(0,0)$. Note: The total field in the forward direction is the sum of the incident and scattered fields. Further, there will be an infinite sum of evanescent (exponentially decreasing) waves. They make up the near field associated with the array.

---

*In the following, the first superscript refers to the array number, the second to the element section.

Denoting the element current in column q and row m by $I_{q,m}^{1,1}$, it follows from Floquet's theorem [5] that the element currents are given by

$$I_{q,m}^{1,1} = I_{0,0}^{1,1} e^{-j\beta q D_x s_x} e^{-j\beta m D_z s_z} \tag{1.3}$$

(i.e., they all have the same amplitude, and a phase which matches that of the incident plane wave with direction of propagation $\hat{s}$).

It has been shown rigorously that the electromagnetic fields from an infinite array are given by a spectrum $\hat{r}_\pm$ of inhomogeneous plane waves [5,24–27]:

$$d\bar{H}^{1,1} = I_{0,0}^{1,1} dl^{1,1} \frac{1}{2D_x D_z} \sum_{k=-\infty}^{\infty} \sum_{n=-\infty}^{\infty} \frac{e^{-j\beta(\bar{R}-\bar{R}^{1,1})\cdot\hat{r}_\pm}}{r_y} [\hat{p}^{1,1} \times \hat{r}_\pm]$$

$$\text{for } y \gtrless 0 \tag{1.4}$$

$$d\bar{E}^{1,1} = I_{0,0}^{1,1} dl^{1,1} \frac{Z}{2D_x D_z} \sum_{k=-\infty}^{\infty} \sum_{n=-\infty}^{\infty} \frac{e^{-j\beta(\bar{R}-\bar{R}^{1,1})\cdot\hat{r}_\pm}}{r_y} [\hat{p}^{1,1} \times \hat{r}_\pm] \times \hat{r}_\pm$$

$$\text{for } y \gtrless 0 \tag{1.5}$$

The spectrum $\hat{r}_\pm$ denotes the directions of the inhomogeneous plane waves emanating from the array. They are found to be [5]

$$\hat{r}_\pm = \hat{x} r_x + \hat{y} r_y + \hat{z} r_z$$

$$= \hat{x}\left(s_x + k\frac{\lambda}{D_x}\right) \pm \hat{y} r_y + \hat{z}\left(s_z + n\frac{\lambda}{D_z}\right) \quad \text{for } y \gtrless 0 \tag{1.6}$$

where

$$r_y = \sqrt{1 - (s_x + k\frac{\lambda}{D_x})^2 - (s_z + n\frac{\lambda}{D_z})^2} \tag{1.7}$$

The fields expressed by equations (1.4) and (1.5) depend on r_y as follows: For the principal direction $k, n = 0,0$, we see from (1.7) that r_y is always real since $|s_x|, |s_z| \leq 1$ [see (1.2)]. This corresponds to a plane wave $\hat{r}_+(0, 0)$, transmitted in the forward direction $\hat{s}$ and another reflected in the specular direction $\hat{s}_s = \hat{r}_-(0, 0) = \hat{x} s_x - \hat{y} s_y + \hat{z} s_z$, as illustrated in Figure 1.10. For $|k|, |n| > 0, 0$, r_y may still be real provided that the interelement spacings D_x and D_z are large enough. These directions are termed *grating lobe directions*. They are discussed in Section 1.6.4 (see also Section 1.5).

However, for higher values of k and n, r_y will always be imaginary; that is, the exponent in the plane waves $e^{-j\beta(\bar{R}-\bar{R}^{1,1})\cdot\hat{r}_\pm}$ will be real, depicting evanescent waves that go to zero as the point of observation $\bar{R}$ moves away from the array, as illustrated in Figure 1.12. (Formally, it is, of course, possible to choose the sign for r_y such that the evanescent field components would increase exponentially to infinity as we move away from the array. However, such a solution is obviously invalid since it violates fundamental physical laws.) The sum of these evanescent waves constitutes the near field surrounding the elements.

Note: Our array is located in an ordinary dispersionless media and not in Veselago's medium. Also, the field vectors $d\bar{E}^{1,1}$ and $d\bar{H}^{1,1}$ are oriented along $[\hat{p}^{1,1} \times \hat{r}_\pm] \times \hat{r}_\pm$ and $[\hat{p}^{1,1} \times \hat{r}_\pm]$, respectively (i.e., $d\bar{E}^{1,1}$ and $d\bar{H}^{1,1}$ and the propagation $\hat{r}_\pm$ form a right-handed triplet). It is relatively simple to show that it holds as well when r_y becomes imaginary (i.e., for the evanescent waves).

Also, if the array is located in a dispersionless medium, Poynting's vector will coincide with the directions of propagation $\hat{r}_\pm$ as given by equations (1.6) and (1.7). Thus, the spectrum of plane waves radiated from this simple periodic structure will definitely be right-handed and never left-handed as is the case for "Veselago's medium."

1.6.2 Single Array of Elements with Two Segments

Next, we again consider a single array, but this time with elements made of two segments with arbitrary orientation $\hat{p}^{1,1}$ and $\hat{p}^{1,2}$, elements length $dl^{1,1}$ and $dl^{1,2}$, currents $I^{1,1}$ and $I^{1,2}$, and reference points $\bar{R}^{1,1}$ and $\bar{R}^{1,2}$, respectively. Obviously, the array with element orientation $\hat{p}^{1,2}$ has the same interelement spacings D_x and D_z as the first [i.e., the two arrays have the same spectrum $\hat{r}_\pm$; see equations (1.6) and (1.7)]. Thus, the fields from the array with orientation $\hat{p}^{1,2}$ are

$$d\bar{H}^{1,2} = I_{0,0}^{1,2} dl^{1,2} \frac{1}{2D_x D_z} \sum_{k=-\infty}^{\infty} \sum_{n=-\infty}^{\infty} \frac{e^{-j\beta(\bar{R}-\bar{R}^{1,2})\cdot\hat{r}_\pm}}{r_y} (\hat{p}^{1,2} \times \hat{r}_\pm)$$

$$\text{for } y \gtrless 0 \tag{1.8}$$

$$d\bar{E}^{1,2} = I_{0,0}^{1,2} dl^{1,2} \frac{Z}{2D_x D_z} \sum_{k=-\infty}^{\infty} \sum_{n=-\infty}^{\infty} \frac{e^{-j\beta(\bar{R}-\bar{R}^{1,2})\cdot\hat{r}_\pm}}{r_y} (\hat{p}^{1,2} \times \hat{r}_\pm) \times \hat{r}_\pm$$

$$\text{for } y \gtrless 0 \tag{1.9}$$

The total H-field from the combined array is obtained by addition of equations (1.4) and (1.8):

$$d\bar{H} = d\bar{H}^{1,1} + d\bar{H}^{1,2}$$

$$= \frac{1}{2D_x D_z} \sum_{k=-\infty}^{\infty} \sum_{n=-\infty}^{\infty} \frac{e^{-j\beta\bar{R}\cdot\hat{r}_\pm}}{r_y} (\hat{p}^{1,1} dl^{1,1} I^{1,1} e^{j\beta\bar{R}^{1,1}\cdot\hat{r}_\pm}$$

$$+ \hat{p}^{1,2} dl^{1,2} I^{1,2} e^{j\beta\bar{R}^{1,2}\cdot\hat{r}_\pm}) \times \hat{r}_\pm \qquad \text{for } y \gtrless 0 \qquad (1.10)$$

Similarly, the total E-field from the combined array is obtained by addition of equations (1.5) and (1.9):

$$d\bar{E} = d\bar{E}^{1,1} + d\bar{E}^{1,2}$$

$$= \frac{Z}{2D_x D_z} \sum_{k=-\infty}^{\infty} \sum_{n=-\infty}^{\infty} \frac{e^{-j\beta\bar{R}\cdot\hat{r}_\pm}}{r_y} [(\hat{p}^{1,1} dl^{1,1} I^{1,1} e^{j\beta\bar{R}^{1,1}\cdot\hat{r}_\pm}$$

$$+ \hat{p}^{1,2} dl^{1,2} I^{1,2} e^{j\beta\bar{R}^{1,2}\cdot\hat{r}_\pm}) \times \hat{r}_\pm] \times \hat{r}_\pm \qquad \text{for } y \gtrless 0 \quad (1.11)$$

Inspection of equations (1.10) and (1.11) shows readily that a single array with elements comprised of two segments will have a field where $d\bar{E}$, $d\bar{H}$, and $\hat{r}_\pm$ form a right-handed system.

Note: There will, in general, be strong coupling between the two segmented arrays such that $I^{1,1}$ and $I^{1,2}$ may differ significantly from the single-segment cases. This coupling is incorporated in our theory and the PMM program* such that the array currents are always calculated correctly.

1.6.3   Single Array of Elements with an Arbitrary Number of Segments

Extension from two to an arbitrary number of element segments is done simply by induction. Again, we conclude that only right-handed waves will emanate from a single array, regardless of the shape of the elements.

---

*PMM stands for *periodic method of moments*. It is available from the U.S. Air Force. It was written by Lee Henderson as part of his dissertation at the Ohio State University. It is considered one of the fastest and most reliable programs available.

1.6.4   On Grating Lobes and Backward-Traveling Waves

When r_y is real, we experience propagating plane waves. We saw earlier that we always have two propagating waves for k, $n = 0,0$, corresponding to the forward and reflected waves shown in Figure 1.12 (these are also called the *principal waves*). However, as seen by inspection of equation (1.7), we may also obtain propagating waves for a limited number of values of k, n, depending on the interelement spacings D_x and D_z as well as s_x and s_z. The lowest-order grating lobe is obtained for either $s_z = 0$ with k, $n = -1, 0$ or $s_x = 0$ with k, $n = 0, -1$. The latter case is illustrated in Figure 1.13. Note that the component of the phase velocity along the z-direction is opposite for the incident and lowest grating lobe directions. For that reason this grating lobe has sometimes mistakenly been denoted as a *backward-traveling wave*. These grating lobes are encountered in numerous microwaves devices, such as the backward-traveling oscillator, the backward-traveling antenna, and photonic bandgap materials. They have long been known and are well understood. Again, we emphasize that these backward-traveling waves can exist only when the interelement spacings D_x and D_z exceed $\lambda/2$.

The term *backward-traveling wave* was later suggested to mean a wave where the phase and group velocities were opposite each other [28–31].

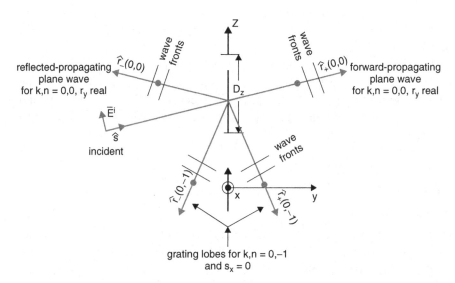

Figure 1.13 Plane wave incident upon an infinite array may also, in addition to the forward and specular reflected waves, produce plane waves in the grating lobe direction $\hat{r}_+(0, -1)$ and $\hat{r}_+(0, -1)$ if the interelement spacings $D_z > \lambda/2$ and $s_x = 0$. Note: All waves are right-handed.

It was thought that Veselago associated such waves with his findings for media with negative μ and ε when he concluded that the phase velocity and the Poynting vector were opposite each other. However, this writer is not aware that he ever used the term *backward-traveling wave*. It should be emphasized that Veselago and his followers, in general, consider media with interelement spacings of less than $\lambda/4$ (denoted *continuous*); that is, we are definitely not talking about grating lobes here. Furthermore, these newer backward-traveling waves can only exist in a highly dispersive medium. It is claimed that it is possible to construct these artificially by periodic loading of a transmission line [28–31] (see also Section 1.9). This writer is not aware that the equivalent was ever done in free space. At any rate, group velocity and phase velocity are the same for free space and a dispersionless medium. In other words, there is absolutely nothing "backward" about any of the plane waves emanating from a periodic structure in Figure 1.13 as long as it is placed in a medium without dispersion. The Poynting vector for all these plane waves points in the direction of propagation $\hat{r}_\pm$ regardless of the number of element segments or element shapes.

1.6.5 Two Arrays of Elements with an Arbitrary Number of Segments

So far we have considered only a single array with an arbitrary number of element segments. We found that the field emanating from such an infinite array consisted of a spectrum $\hat{r}_\pm$ of inhomogeneous plane waves, as given by equations (1.6) and (1.7):

1. A propagating wave in the forward and specular directions corresponding to $k, n = 0,0$ (also called the *principal directions*)
2. A finite number of grating lobes if the interelement spacings D_x and D_z are large enough, corresponding to a finite number of $k, n \neq 0,0$
3. An infinite number of evanescent waves that go to zero as we move away from the array

As shown earlier, all of these waves are right-handed. We now place another array a certain distance d_1 to the right of the first array, as illustrated in Figure 1.14. The interelement spacings D_x and D_z are the same as for array 1, but the number of element segments is arbitrary. Thus, the spectrum $\hat{r}_\pm$ is the same for the two arrays.

We now calculate the currents in all the element segments. Just as the coupling between the segments in one array can be significant, as noted above, it will also be significant between the segments in the two

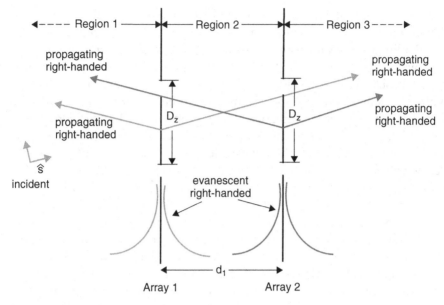

Figure 1.14 Plane wave with direction of propagation $\hat{s}$ incident upon two arrays with interelement spacings D_x and D_z. Each array emanates plane propagating waves in the forward as well as the specular directions; they are all right-handed and so are their sums, regardless of region. Further, there will be an infinite sum of evanescent (exponentially decreasing) waves that represent the near field associated with both arrays. Note: The arrays are located in a medium without dispersion.

arrays. We emphasize that this coupling is always taken rigorously into account both in the theory treated in refs. 5,24, and 25 and in the PMM program [26,27]. Once we find all the segment currents in both arrays in each other's presence, the determination of the fields emanating from each array is done precisely as was done for the single-array case treated earlier and as shown in Figure 1.14. We define three regions:

- Region 1 is the semi-infinite space to the left of array 1.
- Region 2 is the space between arrays 1 and 2.
- Region 3 is the semi-infinite space to the right of array 2.

In region 1 we observe left-going propagating waves radiating from the two arrays; similarly, we have right-going waves in region 3; and we have both left- and right-going waves in region 2, as shown. All of these waves are right-handed. The total field is obtained simply by superposition of the fields from the two arrays. There can be no doubt that in regions 1 and 3 the total field will be right-handed. Further, in region 2 we simply obtain a total field of two right-handed waves crossing each other. Neither one of these waves can ever turn into left-handed waves, since that would require the presence of Veselago's magic material, which everyone agrees does not exist in nature. Remember that our support medium is assumed to have no dispersion (i.e., linear).

1.6.6   Can Arrays of Wires Ever Change the Direction of the Incident Field?

Even for a multilayer, infinite array with identical interelement spacings of less than $\lambda/2$, the element currents in each array will always follow Floquet's theorem [see (1.3)]. As shown earlier, this can only lead to a plane-wave spectrum with directions $\hat{r}_\pm$ (i.e., never "bend" the incident field unless the waves are somehow slowed down). Molecular "dipoles" are a different matter (see also Sections 1.12.1 and 1.12.2 regarding artificial dielectrics.

1.7   ON INCREASING EVANESCENT WAVES: A FATAL MISCONCEPTION

The total evanescent field in Figure 1.14 is obtained by superposition of the evanescent waves from each array. However, these will, in general, not be in phase, and thus the total field cannot be obtained by simple addition of the magnitudes from the individual arrays. In fact, they could be out of phase and actually produce a null somewhere between the two arrays. Whatever the case, it is obvious from inspection of Figure 1.14 that the total field can increase only when the point of observation moves close to the elements, not all of a sudden because we are in a "Veselago medium." We are still in a medium without dispersion, and straightforward rules prevail. This writer is not aware of any demonstration of increasing evanescent waves except on capacitively loaded transmission-line models terminated in a resistive load [28,29].

It is, of course, quite possible to have a multiarray configuration as shown in Figure 1.15 or a transmission line where the last array has a much stronger current than that of the other arrays. (This situation could easily be obtained by loading the arrays in front of the last array either resistively or reactively.) Obviously, the total field will be dominated by the field from the last array, and this situation could be misinterpreted as an "evanescent" wave that "grows" as it moves through some "magic" material. Remember, you are in ordinary air between the elements where the classical laws of electromagnetics prevail.

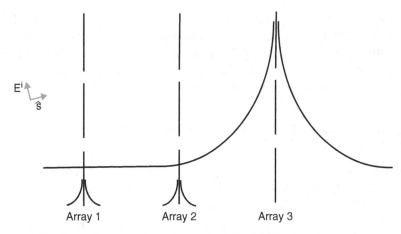

Figure 1.15 Several arrays where the one to the right is designed to have a much stronger element current than the others. This will produce a dominating evanescent field that often is misinterpreted as "proof" that the evanescent wave(s) can increase as you move from left to right.

1.8   PRELIMINARY CONCLUSION: SYNTHESIZING VESELAGO'S MEDIUM BY A PERIODIC STRUCTURE IS NOT FEASIBLE

In Section 1.2 we presented what is widely believed about Veselago's medium. We emphasize that the negative refraction was conceived by Veselago, but some of the other features were suggested by others. Although such materials have never been found in nature, it was suggested by Pendry that such materials could be synthesized by periodic structures with special elements. However, we found some troubling deviation between Veselago's theoretical material and what can de facto be obtained by using periodic structures. Regardless of element shape, the most prevalent factors were:

1. A negative index of refraction is observed between Veselago's medium and a medium with ε, $\mu > 0$. The phase match between the incident and refracted fields was explained by the concept of backward-traveling waves, as discussed in refs. 28–31. However, no trace of such waves was found in a lossless periodic structure, although they can exist on cables terminated in a proper load, as explained in Section 1.9. Experimental evidence of negative refracted fields in a finite periodic structure is plagued by persistent unexplained loss in excess of about -14 to 20 dB [10,14]. This writer has suggested that the field

observed is not a refracted field but radiation from a surface wave characteristic of finite periodic surfaces [20,21]. Further, we found no evidence that periodic structures with interelement spacings of less than λ/2 could change the direction of the incident field, as one would expect for an index of refraction $n \neq 1$ (however, see also Section 1.12.2).

2. It is widely believed that the input impedance of Veselago's medium mounted in front of a ground plane can rotate the "wrong" way (counterclockwise) in the Smith chart (see Figure 1.2 and refs. 2–4). We found absolutely no indication of such a phenomenon in lossless periodic structures suspended in a dispersionless medium (however, see also the discussion in Section 1.9).

3. Just as propagating waves in Veselago's medium can rotate the "wrong" way in a Smith chart, it is quite logical that evanescent waves might increase. In fact, it is generally believed that Veselago's material will support an evanescent wave that increases as you move away from the source (see, e.g., ref. 28, Fig. 3.27 and Sec. 3.7). We found that a periodic structure could only produce truly evanescent waves that would decrease as you move away from the individual arrays. Surely, a multiarray configuration could be designed such that a superficial look could give the impression that an evanescent wave increases as you go through the periodic structure (see Figure 1.15).

4. Veselago claims that a plane wave propagating through his material is left-handed; that is, $\bar{E}, \bar{H}$, and the direction of propagation (phase) form a left-handed triplet, while $\bar{E}, \bar{H}$, and Poynting's vector (energy direction) form a right-handed triplet as usual, regardless of the handedness of the medium. This implies that we will observe a time advance as we move away from the source (see Figure 1.2 as well as ref. 2). This concept is explained alternatively by backward-traveling waves [30,31]. (Note that very few of the classical textbooks treat this subject at all.)

However, we found from rigorous calculations that the field from an infinite periodic structure regardless of the element shape is always right-handed, both inside and outside the periodic structure. Further, there was never any trace of backward waves whatsoever. And as all experienced antenna engineers know, nothing ever moves backward in a Smith chart as long as our load impedance is purely imaginary (Foster's reactance theorem).

It should finally be emphasized that all impedance components in the discussion so far have been completely lossless, including the termination of the space behind the periodic structure. When resistive or dielectric loss is present, the situation changes radically, even if only the termination is lossy. Basically we will, in that case, move inside the rim of the Smith chart such that Foster's reactance theorem no longer holds. This case is discussed in the next section, where we illustrate a typical case in the form of a transmission line terminated in a complex load. This is a little easier than a periodic structure to comprehend, and it has already been discussed in several places [28,29]. Subsequent extension to periodic structures will be facilitated (see Section 1.9.2).

1.9   ON TRANSMISSION-LINE DISPERSION: BACKWARD-TRAVELING WAVES

1.9.1   Transmission Lines

One of the most remarkable conclusions above was that the input impedance of a lossless transmission line terminated in a pure reactance is always located on the rim of the Smith chart and always runs clockwise with frequency (see Figure 1.2), never the other way around unless you *really* have a negative index of refraction. But what if the transmission line is terminated in a complex load rather than a pure reactance?

In fact, this problem has been investigated in numerous papers and at least four books [28–31]. The approach taken there is to start with the equivalent circuit for an ordinary transmission line (i.e., comprised of series inductors and parallel capacitors). The next step is to use duality to obtain an equivalent circuit with series capacitors and parallel inductors. By using simple first-order approximations, it is shown next that the dual circuit has a phase velocity equal to the negative of its group velocity. We shall not repeat the derivation here, since it suffers from several flaws, one being that the result is incorrect, and another that we end up with a dual circuit without a transmission line. This "essential" part could certainly be added later, but that approach leads to unnecessary complications and is still not satisfactory [28].

It is, in fact, usually much better to ask a direct question: What can be done to eliminate or at least reduce the dispersion of a transmission line? Actually, it has very little to do with duality. In fact, this problem is solved in the most direct way by use of the Smith chart, as illustrated by the following example.*

---

*The Smith chart is often frowned upon as being an approximate graphical approach. However, we should hasten to emphasize that the Smith chart represents a graphical illustration of an *exact* solution, not just some first-order approximate formulas. Most important, it depicts exactly what goes on in the complex plane and helps us in our thought process.

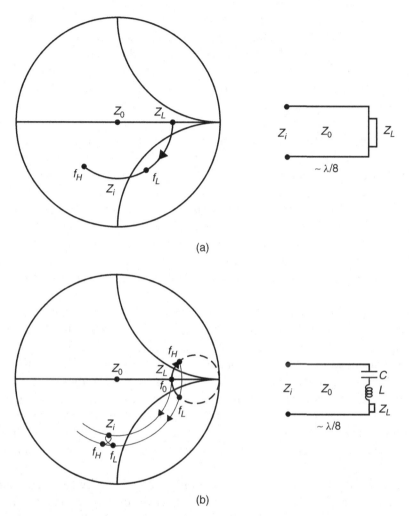

Figure 1.16 (a) Smith chart showing the input impedance Z as a function of frequency of a transmission line terminated in a load impedance Z_L (real). (b) The same as in part (a) above but with an LC series circuit to Z_L. It is seen to reduce the dispersion of Z_i; in fact, it does run backward over a limited frequency range. This does not violate Foster's reactance theorem because Z_L is lossy, so we are not on the rim of the Smith chart.

In the insert of Figure 1.16a we show a transmission line with characteristic impedance Z_0, length $\sim \lambda/8$, and terminated with a load impedance $Z_L \sim 2Z_0$. The input impedance Z_i as a function of frequency will typically look as shown in the Smith chart in the same figure, where the gap between the low frequency f_L and the high frequency f_H is an indication of the dispersion of the transmission line.

We ask a simple question: Is there any way in which this dispersion gap can be reduced or perhaps even reversed? As we shall see, there is indeed, but only if Z_L is located inside the Smith chart (i.e., has a resistive component), never when it is located on the rim of the Smith chart (i.e., is purely reactive). We illustrate this statement by the example shown in Figure 1.16b. As seen in the insert, we have added a series LC circuit to $Z_L \sim 2Z_0$ that resonates at the center frequency, $f_0 \sim \frac{1}{2}(f_L + f_H)$. In other words, the effective load impedance for the transmission line will be located on part of a circle going through Z_L and the infinity point, as shown (see Appendix B of ref. 32). Note that this impedance, when seen from the center of the Smith chart, will rotate counterclockwise (the "wrong" way) as the frequency increases from the low frequency f_L to the high frequency f_H. In other words, if we next add the clockwise rotation from the transmission line, we obtain an input impedance, Z_i, with a strongly reduced gap between f_L and f_H (i.e., we have reduced the dispersion for the transmission line and part of the curve is actually running "backward").

This approach can be extended and modified in many ways. If, for example, we extend the length of the cable from about $\lambda/8$ to about $\lambda/4$, as shown in Figure 1.17a, the new Z_i may have the high frequency f_H, somewhat ahead of the low frequency f_L (i.e., we see a moderate dispersion). However, if we note that the impedance of a parallel LC circuit is located on the rim of the Smith chart around the infinity point of the Smith chart, as shown, it is easy to see that adding this impedance in parallel with Z_i will result in a new Z_i where the dispersion even for this longer cable is strongly reduced, as shown in Figure 1.17b (see Appendix B in ref. 20).

We can extend this approach indefinitely, alternating between series and parallel LC circuits. It is easy to see that the waves on this composite cable can be considered as a combination of forward- and backward-traveling waves, where the first is always present and the relative strength of the second depends on the specific design. Although this is all well and good, it would be erroneous to think that we have produced a new exotic material. In fact, if we let the original load impedance, Z_L, go toward the rim of the Smith chart, we observe that only the forward-traveling wave will remain. Or put another way, if we cut a section out of our composite cable, it has no particular redeeming feature. We have simply demonstrated some old network tricks, well known for broadband matching technique [20,32].

It should finally be noted that the concept as presented here has some similarities with the circuit obtained by the duality concept: for example, the use of parallel capacitors. However, it fails to use the inductors, which give a greater variation with frequency. Also, it does not alternate between parallel and series LC circuits at every $\lambda/4$ separation. All in all, the duality approach is lacking compared to the circuit presented here. Actually, there is very little justification in using duality to deal with this problem.

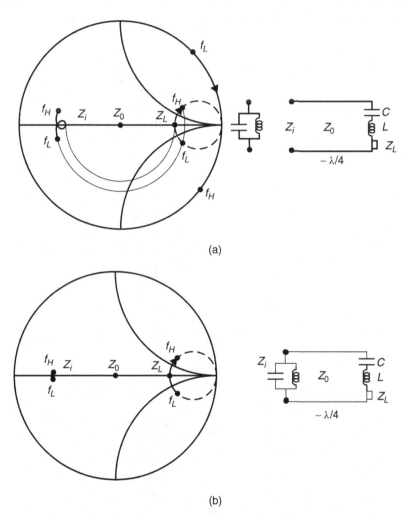

(a)

(b)

Figure 1.17 (a) The same as in Figure 1.16a but for a longer transmission line. Also indicated to the right in the Smith chart is the impedance of a parallel LC circuit. (b) When the parallel LC circuit is added to the left of the transmission line, we observe reduced dispersion of Z_i, even backward-traveling waves over a limited frequency range. Note: We have not "invented" a new "material," since it falls apart for Z_L reactive and other cases as well.

1.9.2 Periodic Structures

We investigated above the possibility of limiting or even reversing dispersion on a transmission line. This background will greatly facilitate our extension to periodic structures. An example is shown in Figure 1.18a, where we show a slotted frequency-selective surface (FSS) to the right and a dipole FSS to the left. This case differs from the transmission-line case in Figures 1.16 and 1.17 by the fact that the space to the right with intrinsic impedance Z_0 will put us right in the center of the Smith chart

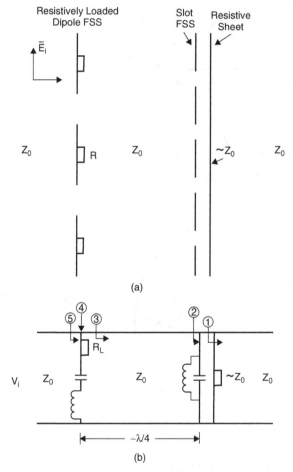

Figure 1.18 (a) Actual configuration of a combination of a slot FSS backed by a resistive sheet with a sheet resistance of about Z_0. To the left is a resistively loaded dipole FSS. The incident field is coming from the left as shown. (b) Equivalent circuit of the configuration shown in part (a). The circled numbers refer to the impedances looking to the right except that ④ refers simply to the loaded dipole FSS without space behind it.

and not at about $2Z_0$, as shown in Figure 1.16. The remedy for this dilemma is simply to add a resistance sheet of about Z_0 in parallel with the space impedance as shown in Figure 1.18a and also in the equivalent circuit in Figure 1.18b. This results in a total resistance ① equal to about $Z_0/2$. We now add the slotted FSS in parallel. Recalling that the equivalent circuit for a slotted FSS is a parallel LC circuit, we readily see that the total impedance ② looking to the right is merely located on a circle going through zero and about $Z_0/2$, as indicated in the Smith chart in Figure 1.19a. Note that when seen from the center of the Smith chart, this impedance curve runs the "wrong" way (counterclockwise). Thus, the impedance ③ obtained by clockwise rotation of ② has reduced dispersion.

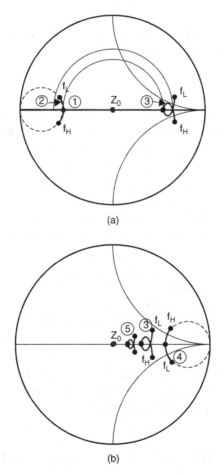

(a)

(b)

Figure 1.19 (a) The various impedances as denoted in Figure 1.16b shown in a Smith chart up to ③. (b) The remaining impedances from Figure 1.16b shown in a Smith chart.

Further reduction is obtained by adding a loaded dipole FSS. We recall that the equivalent circuit ④ for this configuration is a series RLC circuit, as shown in Figure 1.18b. As shown in the Smith chart in Figure 1.19b, this impedance is located on a circle going through R_L and the infinity point. Again we observe that the impedance ④ runs the "wrong" way (i.e., it will reduce the dispersion of impedance ③).

However, we see no particular reason at this point to continue this discussion concerning the possibility of creating new "materials" with a negative index of refraction. As mentioned earlier, we have really not created any new unique medium but merely applied a well-known approach from broadband matching techniques [20,32]. And it is far from lossless. Thus, you can forget about amplification of the evanescent waves.

1.10   REGARDING VESELAGO'S CONCLUSION: ARE THERE DEFICIENCIES?

1.10.1   Background

In 1968, Veselago asked a simple question: What would happen if both μ and ε for a material were negative [1]? He concluded that the index of refraction, n_1, between an ordinary medium with μ, ε, > 0 and one with μ, $\varepsilon < 0$ would be negative. Further, while Poynting's vector would propagate in the usual direction, the phase vector would point backward, later giving rise to the term *backward-traveling waves*. More extensions of Veselago's conclusions were added later by others. However, Veselago had conceived the most important aspect: a negative index of refraction. He stated quite correctly that no such material had ever been found or produced, and he very prudently added that there were, perhaps, very good reasons for the absence of such materials.

It was eventually suggested by Pendry almost 30 years later that materials with μ, $\varepsilon < 0$ could be produced artificially by a periodic structure comprised of special elements [6–9]. We investigated that possibility earlier and concluded that *none* of the features characteristic of Veselago's medium could be produced by a periodic structure regardless of the type of element. Given that fact, it is natural to ask the simple question: Is Veselago's medium physically realizable?

1.10.2   Veselago's Argument for a Negative Index of Refraction

Veselago arrived at his conclusions by considering the boundary conditions between two media, 1 and 2, as shown in Figure 1.20. He first stated that the tangential components for the two media must be equal regardless of the sign of μ and ε in the two media; that is (using Veselago's notation),

$$E_{t1} = E_{t2} \qquad H_{t1} = H_{t2} \tag{1.12}$$

Further, the boundary conditions for the normal components states that

$$\varepsilon_1 E_{n1} = \varepsilon_2 E_{n2} \qquad \mu_1 H_{n1} = \mu_2 H_{n2} \tag{1.13}$$

Thus, we see clearly that if ε_1, μ_1 and ε_2, μ_2 have the same signs, the direction of propagation $\bar{k}_2$ in medium 2 will be as indicated in Figure 1.20 for $n_{12} > 0$. However, if ε_2, μ_2 has the sign opposite that of ε_1, μ_1, the normal components of $\bar{E}$ and $\bar{H}$ will be opposite each other according to (1.13), which means that the phase velocity k_{v2} in medium 2 will be as indicated in Figure 1.20: left-handed. However, we also note that

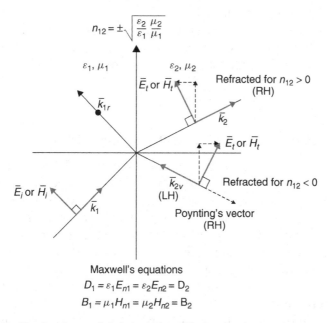

Figure 1.20   Veselago's proof that the index of refraction is negative for two media if ε_1, $\mu_2 > 0$ and ε_2, $\mu_2 < 0$. His argument is that the tangential field components must be the same regardless of handedness. However, the normal components change sign with $\varepsilon_1/\varepsilon_2$ and μ_1/μ_2 according to Maxwell's equations, as indicated in the figure. His proof is correct, but only mathematically, since it implies negative propagation constant β_2 and ultimately negative time (see the discussion related to Figure 1.21).

Poynting's vector (in cgs units) is given by

$$\bar{S} = \frac{c}{4\pi} \bar{E} \times \bar{H} \qquad (1.14)$$

That is, $\bar{S}$ always forms a right-handed set with the vectors $\bar{E}$ and $\bar{H}$ and will therefore point in the direction opposite $\bar{k}_{2v}$ as also shown in Figure 1.20.

In other words, Veselago had shown that for two media, 1 and 2, where ε_1, μ_1 and ε_2, μ_2 have opposite signs, the index of refraction, n_{12}, would be negative. Furthermore, since the phase delay through a medium is given by $n_{12}\beta_0 d$, we observe immediately that for $n_{12} > 0$ we experience a phase delay and, similarly, a phase advance for $n_{12} < 0$, as illustrated in the Smith chart in Figure 1.2. Such conclusions should immediately raise questions about causality.* Indeed, some papers took issue with Veselago's conclusion, of which the most conspicuous was one by Valanju, Walser, and Valanju [33]. However, that merely led to an exchange of comments from Pendry and others, and eventually died out. In any case, Valanju et al. were never proven wrong. Meanwhile, the stream of papers concerning metamaterials continued unabated, and eventually at least four books on the same subject were published [2,28,29,46].

In this writer's opinion, Walser and associates were right and one may wonder why their paper did not have a greater impact. One reason probably is that it was a little intricate and not immediately understood. Thus, in the following we attempt a simpler explanation and show that Veselago's conclusions have physical deficiencies.

1.10.3   Veselago's Flat Lens: Is It Really Realistic?

The concept for Veselago's flat "lens" is by now well known, as shown in Figure 1.21. It consists of a flat slab where ε_2, μ_2 not only is negative but also $\varepsilon_2 = -\varepsilon_1$ and $\mu_2 = -\mu_1$ (i.e., $n_{12} = -1$) such that the refracted angle, according to Veselago, is always the negative of the angle of incidence. We show two rays emanating from the source point S located to the left. They cross inside the lens at a point denoted cross 1 and outside to the right at a point denoted cross 2. Such crossings are often thought to be focal points. However, more is required for such a classification. Foremost, we must require that all rays arrive with the same phase. Inspection of the two rays show clearly that ray SB is delayed in phase with respect to ray SA_2 by section A_1B. Further, section BA_3 is inside the metamaterials where the signal is advanced precisely by the same amount, according to Figure 1.2, such that the two rays will

---

*After all, how can a signal arrive before it starts?

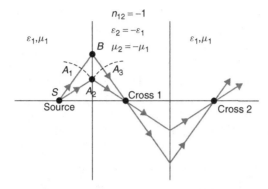

Figure 1.21 Veselago's flat lens with $\varepsilon_2 = -\varepsilon_1$ and $\mu_2 = -\mu_1$. The longest path ray will be delayed in phase corresponding to A_1B but be advanced in Veselago's medium corresponding to BA_3 (see also Figure 1.2). However, if the two rays are to arrive at the same time at cross 1, it must involve negative time in Veselago's medium. See also the discussion in the text in conjunction with equations (1.15) to (1.17) as well as Figures 6.1 and 6.2.

arrive at cross 1 in phase. However, we must also require the two rays to arrive at the crossing *at the same time*. Obviously, that would require the time delay A_1B to be canceled by a time advance BA_3 (i.e., negative time!). Although negative time does not "offend" mathematicians, it is definitely not an option open to physicists, particularly not to engineers.* So no wonder we have trouble synthesizing Veselago's medium!

1.11 CONCLUSIONS

When Veselago published his now famous paper in 1968 [1], he merely asked a simple question: What would happen if both μ and ε were negative? He came up with several interesting conclusions. The most important were:

1. The index of refraction between an ordinary medium and one with μ, $\varepsilon < 0$ would be negative.
2. The field vectors $\bar{E}$ and $\bar{H}$ and the direction of phase propagation would form a left-handed triplet, whereas in an ordinary medium they are right-handed.

*Surely, it is possible for the two rays to arrive at different times and still be in phase, but only for a finite number of discrete frequencies. Thus, it fails for a general modulated signal. Similarly, a "static" case would consist of just one frequency. Since no modulation would be possible in this case, it would be of no practical interest (see also Section 6.5).

Apparently there was little interest in Veselago's work until Pendry in the mid-1990s suggested that material with negative ε and μ could be made artificially by use of periodic structures with special elements [6–9]. He and others subsequently came up with additional conclusions concerning materials with μ, $\varepsilon < 0$. The most important were:

3. Evanescent waves would increase as they propagate through a medium with ε, $\mu < 0$, not decrease as they do in an ordinary medium.
4. The phase would advance in a medium with ε, $\mu < 0$ even if lossless [2–4], not be retarded as in an ordinary medium.

Conclusions 3 and 4 are identical from a mathematical point of view. (The exponent in the phase term goes from imaginary to positive real.) Strangely enough, some can accept one but not the other. (They are, of course, both wrong. The first leads to infinite energy at infinity; the other violates causality. See the comments below in conjunction with equations (1.15) to (1.17).] In this writer's opinion, the first of the conclusions above (i.e., negative index of refraction) has never been demonstrated satisfactorily, despite numerous claims in the literature. Most bothersome is the fact that the "negative refracted" power is always less than about 1 to 2% of the power transmitted through a low-loss dielectric reference material. This writer has suggested that the "refracted" field could simply be radiation from a surface wave characteristic for finite periodic structures with interelement spacings below $\lambda/2$ [20,21]. Or it could be a sidelobe from the mainbeam(s). But it certainly is *not* a refracted field! (See also Appendix D about lossy dielectric wedges.)

The second conclusion, that a material with μ, $\varepsilon < 0$ must have $\bar{E}$, $\bar{H}$, and the propagation factor form a left-handed triplet, is probably the one that will be most difficult to synthesize by an infinite periodic structure. In fact, the field from such a structure was shown rigorously to always be right-handed, regardless of the element type. It would require rewriting Maxwell's equations to come up with a left-handed system!

Similarly, the field from an infinite periodic structure was shown always to consist of either propagating waves with phase retardation as you move away from the structure where they originate, or of evanescent waves that are attenuated as you move away. In other words, as claimed in conclusions 3 and 4, the fields simply could not be synthesized by an infinite periodic structure whether it consisted of a single array or of multiple arrays. Of all these conclusions, 3 and 4 are probably the ones that have been the most difficult for this author to accept. It appears that Pendry was quite comfortable with satisfying pure math and less concerned about physics. It simply makes no physical sense to have the amplitude of the electric field go to infinity as we go toward infinity. Nor can a signal arrive before we send it!

Similarly, having the propagation factor β go negative, as indicated in Figure 1.2, leads to fundamental physical problems. More specifically, let us consider a medium with propagation constant β and phase velocity v. Let us assume further that it will take t seconds to travel a distance of d meters. Then clearly we have

$$d = vt \qquad \text{meters} \tag{1.15}$$

We further have

$$v = \lambda f = \frac{\lambda 2\pi f}{2\pi} = \frac{\omega}{\beta} \tag{1.16}$$

Substituting equation (1.16) into (1.15) yields

$$d = \frac{\omega t}{\beta} \qquad \text{meters} \tag{1.17}$$

Inspection of (1.17) shows that if we assume that the distance d as well as the angular frequency ω are both positive (!), clearly β and t must have the same sign. In particular, if $\beta = n_1\beta_0 < 0$, as shown in Figure 1.2, then clearly time t must be negative as well. This observation supports our discussion in Section 1.10.3 about Veselago's flat lens. It also lends credence to the claim of Valanju et al. [33] that causality is violated for materials with μ, $\varepsilon < 0$ (see also Figures 6.1 and 6.2).

Certainly, Veselago was right when he stated in his original paper that material with μ, $\varepsilon < 0$ has never been found in nature. And he added (very prudently): "There are perhaps good reasons for this." He was, in this writer's opinion, also correct in his proof of negative index of refraction—however, only from a purely mathematical point of view. From a physical point of view, it was deficient because it leads to negative time. Walser et al. saw this very early, in 2002 [33].

This writer attended Engheta's oral presentation in Torino in 2001 [3]. He commented from the floor that he found the paper very interesting but that he did "not believe a word of it because it violated Foster's Reactance Theorem." It was followed by much discussion, but no agreement was reached.

We finally investigated the possibility of backward-traveling waves in transmission lines. These are deemed absolutely essential in obtaining the features characteristic of Veselago's medium. They have been investigated intensely by [28] and [29] using duality. We used a more direct approach here simply by applying a broadband matching technique. We found that it is indeed possible (and well known) to obtain an input impedance of a transmission line terminated resistively that makes a loop running the "wrong" way in the Smith chart, as seen from the center over a limited frequency band. This can be interpreted as a backward-traveling wave superimposed on a forward-traveling wave. But this is possible only if the transmission line is terminated in a resistive load in conjunction with a suitable reactance, never if the load impedance is purely imaginary. In other words, it is possible only when we are inside the Smith chart, where Foster's reactance theorem does not hold. We would therefore not characterize this as a special material (it "works" only when terminated with special loads) but, rather, as an application of the well-known broadband matching technique. And this solution is, of course, inherently lossy.

1.12 COMMON MISCONCEPTIONS

1.12.1 Artificial Dielectrics: Do They Really Refract?

Artificial dielectrics made of arrays of short conducting wires suspended either in free space or in a mother dielectric have been known for more than 50 years. W. E. Kock [34] is usually credited with being the originator of the fundamental idea: that an array of small metallic objects can delay a plane wave propagating through such a medium similar to what is observed in an ordinary dielectric medium compared to free space [34]. It is further believed, at least by some, that this delay can change the direction of propagation.

However, earlier in the chapter we stated categorically that a periodic structure of any conducting planar elements suspended in free space cannot change the direction of a plane wave incident upon such a structure. Obviously we owe the reader an explanation for this discrepancy. We are well aware that we disagree with the prevailing view regarding artificial dielectric.

The concept for artificial dielectric is based on an equivalent transmission line loaded periodically with shunt impedances, Z_s, corresponding to each array as shown in Figure 1.22. It is further well known that for short wires ($2l < 0.3\lambda$) the equivalent shunt impedances Z_s are basically capacitive, resulting in a phase delay compared to that of free space, $\beta_0 d$ per array. This fact is usually taken into account by introducing the effective propagation constant β_{eff}, where in the present case, $\beta_{\text{eff}} > \beta_0$. The theory

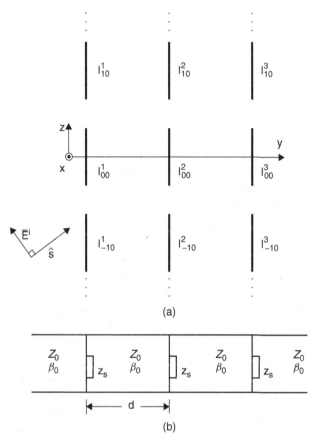

Figure 1.22   (a) Example of an artificial dielectric of small wires suspended in air; (b) equivalent circuit of the artificial dielectric shown in part (a).

for an artificial dielectric now states (defines) that the effective index of refraction is [35,36]

$$n_{\text{eff}} = \frac{\beta_{\text{eff}}}{\beta_0} \qquad (1.18)$$

Certainly, had we considered a homogeneous dielectric material rather than an artificial dielectric of wires, the definition of the index of refraction as given by (1.18) would be correct. However, a more rigorous approach is needed when working with artificial dielectric or periodic structures.

First, we realize that when a plane wave with direction of propagation $\hat{s} = \hat{x}s_x + \hat{y}s_y + \hat{z}s_z$ is incident upon an infinite array in the x- and z-directions, the element currents in column k and row n of array 1 will, according to Floquet's theorem, be given as [5]

$$I_{kn} = I_{00}^1 e^{-j\beta_0 k D_x s_x} e^{-j\beta_0 n D_z s_z} \tag{1.19}$$

Assuming that the interelement spacing D_x, $D_z < \lambda/2$, the element currents given by equation (1.19) will, according to fundamental array theory as shown in Section 1.5, produce propagating plane waves *only* in the forward direction $\hat{s}$ as well as the specular direction $\hat{s}_s = \hat{x} s_{sx} - \hat{y} s_{sy} + \hat{z} s_{sz}$. And the same statement holds for all the other arrays as well (see Chapters 4 and 8 of ref. 5 for details).

Certainly, we hasten to emphasize that the array currents I_{kn}^a, where $a = 1, 2, \ldots, N$, might indeed be delayed or advanced with respect to each other. However, that fact is by itself not capable of changing the *direction* of the radiation from the *individual* arrays. That depends only on the phase distribution across the individual arrays, and that is for an infinite array always given by (1.19) (i.e., Floquet's theorem) and thus will radiate only in the forward direction $\hat{s}$ as well as in the specular direction $\hat{s}_s$. However, see also Section 1.12.2.

If we fill the entire space between the arrays with a material with propagation constant β_1, there will be a change of propagation from $\hat{s}_0$ in air to $\hat{s}_1$ in the "mother" material. It is determined simply by matching phase velocities along the arrays and leads, as is well known from Snell's law. However, there is no additional change of direction due to the periodic structures (for details, see Chapters 4, 5, and 8 in ref. 5). More specifically: The arrays can only affect the propagation constant orthogonal to the arrays, not parallel to them. It appears that only a material with a propagation constant different from that of the incident space, β_0 in this case, can accomplish this.

A very important next step is to break the mother material up into arrays consisting of rectangular "flakes." Such an arrangement of elements that are not simply conducting but have permittivity and/or permeability open up new exciting possibilities, to be treated in a future paper by R. Walser et al.

1.12.2 Real Dielectrics: How Do They Refract?

Actually, what we said earlier about artificial dielectric is only approximately true if the number of parallel arrays is relatively small. In a real dielectric we work with periodic structures where the elements typically are molecular and where the number of arrays is very large indeed. This will result in essentially two things: (1) The delay caused by each array will eventually add up, resulting in a significant change of direction of propagation; and (2) the fields scattered from the individual arrays will add up to a wave propagating in the direction of the incident wave and, eventually, attain equal amplitude and be $180°$ out of phase. This statement is based on the *extinction theorem* presented by Ewald [45]. Thus, the field inside a real dielectric will consist only of a refracted wave propagating in a direction consistent with Snell's law.

There will, of course, be a somewhat similar effect in the artificial dielectric case. However, because the elements typically are larger, the number of arrays tend to be smaller, resulting in a weak effect. Just exactly what constitutes a "small" and a "large" number of arrays is an interesting problem to ponder. Probably the total field transmitted in the forward direction would have an amplitude following a spiral with decreasing radius as the number of arrays increases. It would require extensive computer runs with the PMM code, requiring help from students no longer available. Why? Because when you take on a student, your life expectancy should go beyond about five years, and I am past that limit!

Of course, in an artificial dielectric of small extent, the direction of propagation can change considerably when passing close to the individual element. However, the average direction is the same.

One thing is certain: Neither an artificial nor a real dielectric will produce negative refraction!

1.12.3   On the E- and H-Fields

It was originally suggested by Pendry that ε originates in parallel wires whereas μ is associated with split-ring resonators. It is often implied that the resulting E- and H-fields are independent of each other. This is fundamentally wrong. Only at dc can you control these two field vectors independently. At higher frequencies they become like the two sides of one piece of paper: You cannot have one side without the other. This is a simple consequence of Maxwell's equations.

More specifically, coupling between parallel wires and split-ring resonators is typically assumed to be about zero. Considering that the coupling is actually 100%, it is obvious that this will lead to both computational and conceptual mistakes. What actually takes place in a periodic structure of wires and split-ring resonators is discussed in Appendix A. We do not perform an actual calculation. Rather, to understand what really goes on, we explain the physics behind it, which is more important. Needless to say, we do not observe any negative μ or ε whatsoever.

1.12.4   On Concentric Split-Ring Resonators

The array of split-ring resonators is often made such that a smaller element is mounted concentrically inside a slightly larger one, whether circular or rectangular. The purpose of such an arrangement is, I am told, to obtain a broader bandwidth similar to staggered tuning. This expectation is based on the assumption that the coupling between the concentric elements is zero, or at least "not important." Nothing could be further from the truth. In fact, the coupling between concentric arrays is 100%. We do obtain two resonances, but instead of having a small valley between them, we find that they are separated by an infinite deep null (assuming that interelement spacing is small enough not to have grating lobes). Double tuning of arrays in general is discussed in detail in Chapter 9 of ref. 20.

1.12.5   What Would Veselago Have Asked if . . .

When Veselago asked his famous question in 1968 [1], he was obviously envisioning Maxwell's equations written in the usual form using μ and ε. However, as shown in Appendix A, it is also possible to write Maxwell's equations in a form that does not contain μ and ε, but instead, the propagation constant $\beta = \omega\sqrt{\mu\varepsilon}$ and the intrinsic impedance $Z = \sqrt{\mu/\varepsilon}$.* This makes quite a bit of sense since we typically measure β and Z from where we obtain μ and ε. Also, we are in general, from both a theoretical and a practical point of view, more interested in β and Z than in μ and ε (see Appendix A).

It is quite interesting to speculate what question Veselago would have asked had he used β and Z. Almost everyone agrees that a negative Z makes no sense unless we try to simulate a black hole in outer space. And a negative β could simply indicate a wave propagating in the negative direction but with a phase *delay* as we move away from the source. Although this case would be trivial, it would be quite a different story if a wave propagated with a phase advanced as we moved away from the source. I doubt that Veselago would have fallen into that trap. See also the discussion in connection with equations (1.15) to (1.17) as well as Section 1.5.

Well, Veselago did not use β and Z but μ and ε. And as we all know, his question started almost 30 years later, one of the most controversial subjects in our time. It has resulted in several books and literally thousands

---

*To the best of this writer's knowledge, this was first observed by W. Rotman in 1962 [37]. However, recently it was pointed out by the author's Swedish friend Per Erik Ljung that E. Hallén also considered this subject in his book *Elektricitetslära* (p. 109) as early as 1953.

of papers, but would we have been better off without these? At least I do not think we would be worse off. We have so far not seen any practical use (other than what we could design without any theoretical input from materials with a negative index of refraction).

It is often stated that we do not have the means to make such structures precise and lossless enough today but will perhaps in the future. I do not think so! Anything I have seen so far has been child's play compared to the highly sophisticated structures used in modern technology. No, the problem is pure and simple: The solution just does not exist! I hope this chapter has shed some light on this subject.

1.12.6   On "Magic" Structures

Every so often you see papers that claim a larger transmission through a periodic structure than expected. Typically, we are dealing here with simple structures such as circular holes (or squares, for that matter) in a thin perfectly conducting screen. The claims are based on the assumption that the transmission coefficient is given by the ratio between the sum of the physical area of the holes and the area of the entire screen: in other words, simple physical optics. Apparently, it is not always realized that periodic structures can exhibit resonances. A frequent explanation is based on the presence of a layer of "plasmons" adjacent to the screen. Although such a layer might be a reality at optical frequencies, we have never found it necessary to resort to such a mechanism at microwave frequencies.

True, a periodic structure of circular holes does possess a somewhat peculiar resonance. Actually, a *single* circular aperture does not resonate. What happens in an *array* is simply that just before onset of the lowest-order grating lobe, the lowest-order evanescent mode becomes extremely strong, which manifests itself in a lot of stored energy of such "polarity" that it makes the aperture holes resonate. This layer of stored energy is as close as we get to a "plasmon layer" at microwave frequencies.

Incidentally, any periodic structure with resonances governed primarily by the onset of grating lobes is usually undesirable because these vary so dramatically with frequency and angle of incidence. They were among the first type of periodic structures to be explored more than 40 years ago, and their bad features are well documented [38–40]. I was therefore surprised when I saw an article in *IEEE Transactions on Antennas and Propagation*, [41] in which researchers working in optics had written a paper about periodic structures with circular apertures in the hope that the FSS community would find this new "discovery" useful. For the record, we remind the reader that perfect transmission can be obtained for an array of slots of length about $\lambda/2$ and an arbitrary vanishing narrow slot width provided that the conductivity is 100% (i.e., there is virtually no physical area!) [42–44]—and a myriad of other element types (see Chapter 2 of ref. 5).

REFERENCES

[1] V. G. Veselago, The electrodynamics of substance with simultaneously negative values of ε and μ, *Sov. Phys. Usp.*, vol. 10, no. 4, pp. 509–524, Jan. 1968.

[2] N. Engheta and R. W. Ziolkowski, *Metamaterials: Physics and Engineering Explorations*, IEEE Press/Wiley-Interscience, Hoboken, NJ: 2006.

[3] N. Engheta, Compact cavity resonators using metamaterials with negative permittivity and permeability, *Proceedings on Electromagnetics in Advanced Applications (ICEAA01)*, Torino, Italy, 2001.

[4] N. Engheta, Is Foster's reactance theorem satisfied in double-negative and single-negative media?, *Microwave Opt. Technol. Lett.*, vol. 39, no. 1, pp. 11–14, Oct. 2003.

[5] B. A. Munk, *Frequency Selective Surfaces: Theory and Design*, Wiley, New York, 2000.

[6] J. B. Pendry, Negative refraction makes a perfect lens, *Phys. Rev. Lett.*, vol. 85, no. 18, pp. 3966–3969, Oct. 2000.

[7] J. B. Pendry, A. J. Holden, W. J. Stewart, and I. Youngs, Extremely low frequency plasmons in metallic mesostructure, *Phys. Rev. Lett.*, vol. 76 no. 20, pp. 4773–3776, June 1996.

[8] J. B. Pendry, A. J. Holden, D. J. Robbins, and W. J. Stewart, Low frequency plasmons in thin-wire structures, *J. Phys. Condens. Matter*, vol. 10, pp. 4785–4809, 1998.

[9] J. B. Pendry, A. J. Holden, D. J. Robbins, and W. J. Stewart, Magnetism from conductors and enhanced nonlinear phenomena, *IEEE Trans. Microwave Theory Tech.*, vol. 47, no. 11, pp. 2075–2084, Nov. 1999.

[10] R. A. Shelby, D. R. Smith, and S. Schultz, Experimental verification of a negative refractive index of refraction, *Science*, vol. 292, pp. 77–79, Apr. 2001.

[11] D. R. Smith, W. J. Padilla, D. C. Vier, S. C. Nemat-Nasser, and S. Schultz, Composite medium with simultaneously negative permeability and permittivity, *Phys. Rev. Lett.*, vol. 84, no. 18, pp. 4184–4187, May 2000.

[12] D. R. Smith, D. C. Vier, N. Kroll, and S. Schultz, Direct calculation of permeability and permittivity for left-handed metamaterials, *Appl. Phys. Lett.*, vol. 77 no. 14, pp. 2246–2248, Oct. 2000.

[13] R. A. Shelby, D. R. Smith, S. C. Nemat-Nassser, and S. Schultz, Microwave transmission through two-dimensional, isotropic, left-handed metamaterials, *Appl. Phys. Let.*, vol. 788, no. 4, pp. 489–491, Jan. 2001.

[14] C. G. Parazzoli, R. B. Gregor, K. Li, B. E. C. Koltenbah, and M. Tanielian, Experimental verification and simulation of negative index of refraction using Snell's law, *Phys. Rev. Lett.*, vol. 90, 2003.

[15] J. B. Pryor, Suppression of surface waves on arrays of finite extent, M.Sc. thesis, Ohio State University, 2000.

[16] D. Janning, Surface waves in arrays of finite extent, Ph.D. dissertation, Ohio State University, 2000.

[17] B. A. Munk, Scattering from surface waves on finite FSS, *Proceedings on Electromagnetics in Advanced Applications (ICEAA01)*, Torino, Italy 2001.

[18] B. A. Munk, D. S. Janning, J. B. Pryor, and R. J. Marhefka, Scattering from surface waves on finite FSS, *IEEE Trans. Antennas Propag.*, vol. 49, pp. 1782–1793, Dec. 2001.

[19] D. S. Janning and B. A. Munk, Effect of surface waves on the current of truncated periodic arrays, *IEEE Trans. Antennas Propag.*, vol. 40, pp. 1254–1265, Sept. 2002.

[20] B. A. Munk, *Finite Antenna Arrays and FSS*, Wiley, Hoboken, NJ, 2003.

[21] B. A. Munk, A new interpretation of negative μ_1 and ε_1 produced by a finite periodic structure, *Proceedings on Electromagnetics in Advanced Applications (ICEAA05)*, Torino, Italy, Sept. 2005, pp. 727–732.

[22] L. R. Tan, J. Huangfu, H. Chen, Y. Li, X. Zhang, K. Chen, and J. A. Kong, Microwave solid-state left-handed material with a broad bandwidth and an ultra low loss, *Phys. Rev. B*, vol. 70, 2004.

[23] P. R. Parimi, W. T. Lu, P. Vodo, J. Sokoloff, J. S. Derov, and S. Stridhar, Negative refraction and left-handed electromagnetism in microwave photonic crystals, *Phys. Rev. Lett.*, vol. 92, 2004.

[24] B. A. Munk, *Finite Antenna Arrays and FSS*, Wiley, Hoboken, NJ, 2003, Fig. 4.9.

[25] B. A. Munk, G. A. Burrell, and T. W. Kornbau, A general theory of periodic surfaces in stratified media, *Tech. Rep. 784346–1*, Ohio State University ElectroScience Laboratory, Nov. 1977.

[26] L. W. Henderson, The scattering of planar arrays of arbitrary shaped slot and/or wire elements in a stratified dielectric medium, Ph.D. dissertation, Ohio State University, 1983.

[27] L. W. Henderson, Introduction to PMM, *Tech. Rep. 715582-5*, Ohio State University ElectroScience Laboratory, Feb. 1986.

[28] G. V. Eleftheriades and K. G. Balmain, *Negative-Refraction Metamaterials*, Wiley Interscience, Hoboken, NJ, 2005.

[29] C. Caloz and T. Itoh, *Electromagnetic Metamaterials, Transmission Line Theory and Microwave Applications*, Wiley-Interscience, Hoboken, NJ, 2006.

[30] S. Ramo, J. R. Whinnery, and T. Van Duzer, *Fields and Waves in Communication Electronics*, 3rd ed., Wiley, New York, 1994.

[31] J. A. Kong, *Electromagnetic Wave Theory*, 2nd ed., EMW Pub., Cambridge, MA, 2000.

[32] Radio Research Laboratory, Harvard University, *Very High Frequency Techniques*, McGraw-Hill, New York, 1947, Chap. 3.

[33] P. M. Valanju, R. M. Walser, and A. P. Valanju, Wave refraction in negative-index media: always positive and very inhomogeneous, *Phys. Rev. Lett.*, vol. 88, no. 18, May 2002.

[34] W. E. Kock, Metallic delay lenses, *Bell Syst. Tech. J.*, vol. 27, 1948.

[35] J. Brown, The design of metallic delay dielectrics, *Proc. IEE (London)*, vol. 97, p. III, Jan. 1950.

[36] J. Brown, *Microwave Lenses*, Methuen & Co. Ltd, London, 1953.

[37] W. Rotman, Plasma simulation by artificial dielectric and parallel-plate media, *IRE Trans. Antennas Propag.*, Jan. 1962.

[38] R. B. Kieburtz and A. Ishimaru, Scattering by a periodically apertured conducting screen, *IRE Trans. Antennas Propag.*, vol. 9, pp. 506–514, Nov. 1961.

[39] S. W. Lee, Scattering by dielectric-loaded screen, *IEEE Trans. Antennas Propag.*, vol. 19, pp. 656–665, Sept. 1971.

[40] C. C. Chen, Transmission of microwave through perforated flat plates of finite thickness, *IEEE Trans. Microwave Theory Tech.*, vol. 21, pp. 1–6, Jan. 1973.

[41] M. Beruete, M. Sorolla, I. Campillo, J. S. Dolado, L. Martin-Moreno, J. Bravo-Abad, and F. J. Garcia-Vidal, Enhanced millimeter wave transmission through quasioptical subwavelength perforated plates, *IEEE Trans. Antennas Propag.*, vol. 53, pp. 1897–1903, June 2005.

[42] R. J. Luebbers, Analysis of various periodic slot array geometries using modal matching, Ph.D. dissertation, Ohio State University, 1975.

[43] R. J. Luebbers and B. A. Munk, Analysis of thick rectangular waveguide windows with finite conductivity, *IEEE Trans. Microwave Theory Tech.*, vol. 21, pp. 461–468, July 1973.

[44] R. J. Luebbers and B. A. Munk, Some effects of dielectric loading on periodic slot arrays, *IEEE Trans. Antennas Propag.*, vol. 26, pp. 536–542, July 1978.

[45] P. P. Ewald, The analysis of crystal optics, *Ann. Phys.*, vol. 45, pp. 1–28, 1916.

[46] R. Marques, F. Matrin, and S. Sorolla, *Metamaterials with Negative Parameters: Theory, Design, and Microwave Applications*, Wiley, Hoboken, NJ, 2008.

2 On Cloaks and Reactive Radomes

2.1 CLOAKS

2.1.1 Concept

A new concept has recently been suggested not only to reduce the backscatter of an object but actually to see around it and observe what is behind. More precisely, the incident signal is captured by a cloak surrounding the object, guided around it, and finally escapes out on the back side. Note that we are not absorbing the incident signal, as is usually the case, rather, we are redirecting it away from the backscatter direction. Typically, all this is accomplished by use of metamaterial with negative μ and ε, or so it is claimed.

In this chapter we show that it may at least be possible to capture the incident signal and let it out on the back side. However, we will show that it does not depend on the presence of exotic materials with negative μ and ε, but can be explained by classical electromagnetic theory. Finally, a very simple cloak is suggested merely by using simple FSS structures, leading to potentially better designs.

2.1.2 Prior Art

An example of a cloak is given in *Physical Review*, vol. E75 [1]. A schematic of the concept is shown in Figure 2.1. The target to be "cloaked" consists, in this case, merely of a dielectric cylinder located at the center. The actual cloak is made of flat conducting slabs arranged radially around the dielectric cylinder as shown in the figure. Finally, the spaces between the conducting slabs are filled with dielectrics.

Actual numerical calculations of this design are given in ref. 1 and duplicated in Figure 2.2. The incident wave is coming from the right and going left. We note that the backscattered field, shown in Figure 2.2a, is indeed reduced, while the forward-scattered field is concentrated in a narrow beam. Silverinha et al. [1] explain the workings of this cloak

Metamaterials: Critique and Alternatives, By Ben A. Munk
Copyright © 2009 John Wiley & Sons, Inc.

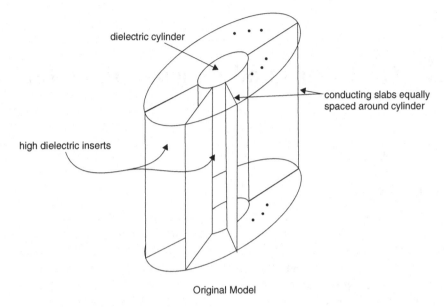

dielectric cylinder

conducting slabs equally
spaced around cylinder

high dielectric inserts

Original Model

Figure 2.1 Cloak comprised of conducting slabs arranged radially around a dielectric
cylinder. The space between the conducting slabs is filled with dielectric. (From ref. 1.)

simply by stating that the flat conductors have an equivalent dielectric
constant less than 1, whereas the dielectric between the conducting slabs
is greater than 1. Thus, it seems to be implied that the "average" dielectric
constant is close to free space, making possible entrance into the space
between cloak and object.

2.1.3 Alternative Explanation

However, a straightforward and correct explanation that does not depend
on materials with negative μ and/or ε is obtained by inspection of the
equivalent circuit shown in Figure 2.3. We first note that the conductive
slabs arranged around the center core are equivalent to shunt inductors.
(This fact has been known for a very long time; see, for example, the 1919
patent of Marconi and Franklin [2, pp. 7–8]). Also, direct calculation by
use of the well-tested PMM program yields an inductance unless grating
lobes are present and interelement spacings are larger than 0.5λ and for
an oblique angle of incidence. Further, the dielectric segments located
between the conducting slabs are equivalent to capacitors in parallel with
the equivalent inductors, as shown in Figure 2.3. Thus, around the resonant
frequency the shunt impedance seen by an incident wave becomes very
large, which enables the signal to penetrate the centerpiece and eventually

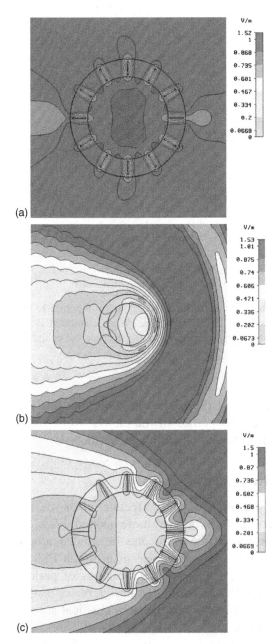

Figure 2.2 Field calculated for the cloak shown in Figure 2.1. (a) The incident field from the right goes through the cloak without significant backscatter; (b) and (c) show the effects of modifying the slabs. (From ref. 1, with permission of Nader Engheta.)

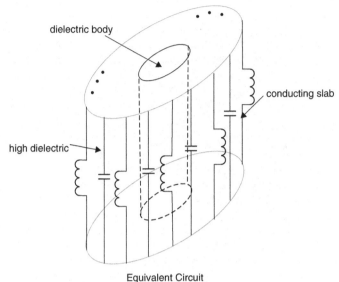

Equivalent Circuit
Note: No Negative μ or ε Whatsoever!

Figure 2.3 Equivalent circuit for the cloak shown in Figure 2.1. The conducting slabs are inductive, the dielectric inserts capacitive. When connected in parallel they form a high impedance, enabling the incident wave to penetrate the cylinder and be re-radiated in the back.

go around and escape at the back side. Note that this explanation relies only on the physical properties of the cloak. No negative μ and/or ε is involved, only fundamental electromagnetic theory.

2.1.4   Alternative Design

A simpler and potentially more effective cloak design is shown in Figure 2.4. It consists of one or more slotted FSSs that will let the incident signal inside in the front and be guided around to the back, where it is re-radiated. The advantage of using an FSS design is simply that a considerable knowledge of the workings of periodic structures in general could be put to use during the design phase of this type of cloak. However, whether a cloak is better than a classical absorber in hiding an object is indeed debatable.

2.1.5   What Can You Really Expect from a Cloak?

A cloak is supposed not only to eliminate backscatter as an absorber would do but also to guide the signal around the object and let it escape at the

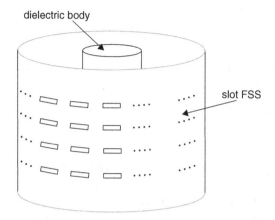

Figure 2.4 Simpler version of the cloak in Figures 2.1 and 2.3, consisting of a slotted FSS wrapped around a dielectric core.

back side. Ultimately, it should also be able to transfer the "back picture" to the front such that you can actually see "through" the object. None of the designs discussed above will do that. In fact, this will require materials that are equivalent to a bundle of optical fibers. So why not use exactly such an arrangement? Materials that will do that will be challenging to produce—but at least they do not violate physical laws.

2.2   REACTIVE RADOMES

Radomes in general are usually designed to be transparent at the operating frequency of the antenna behind them. In this way the antenna impedance undergoes only minor changes. However, if the antenna is mismatched, it is conceptually possible to design the radome such that matching takes place to any impedance desired. We denote such radomes as *reactive* since they act primarily as a lossless tuning device.

2.2.1   Infinite Planar Array with and Without Reactive Radome

The simplest way to illustrate the reactive radome concept is probably in conjunction with an infinite planar array without a ground plane. The equivalent circuit for such a configuration without a radome has already been derived [2, Chaps. 4 and 5] and is shown for easy reference in Figure 2.5. We observe how the antenna terminals are connected via the antenna reactance jX_A to two semi-infinite transmission lines in parallel,

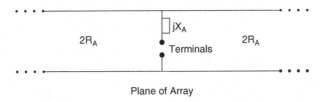

Figure 2.5 Equivalent circuit of an infinite planar array of dipoles without a ground plane. R_A denotes the terminal resistance and jX_A the terminal reactance.

each with characteristic impedance $2R_A$. These transmission lines carry the radiated energy to the left and right of the plane of the array. Note further that R_A denotes the total terminal radiation resistance of the infinite array.

If the total length of the dipole elements is somewhat shorter than $\lambda/2$, the antenna reactance jX_A will typically be negative (i.e., capacitive). In that case we may tune the antenna to resonance simply by inserting an inductance of value $-jX_A$ at the terminals, as shown in Figure 2.6. The radiation resistance R_A may not be exactly the desired terminal resistance but that can eventually be adjusted by use of a transformer or another matching device.

However, instead of using a matching device at the terminals, we may alternatively use a matching device placed suitably somewhere along the two semi-infinite transmission lines. In particular, it could be a dielectric slab that at the same time may serve as a radome. An example of this approach is shown in Figure 2.7. Here we have placed a dielectric slab with intrinsic impedance Z_1 on each side of the infinite dipole array. In the Smith chart of the same figure we show how the radiation resistance $2R_A$ of each semi-infinite transmission line can be transferred by the dielectric slab into the impedance denoted ①, which is chosen such that adding the capacitive antenna reactance, $2jX_A$ in series lands us in the arbitrary resistance $2R_{A1}$ for each side (i.e., a total resistance of R_{A1}). This is accomplished by adjustment of Z_1 as well as the thickness of the slabs.

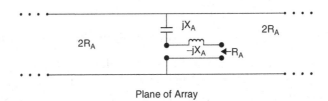

Figure 2.6 If the dipoles are shorter than about $\lambda/2$, jX_A will typically be negative (i.e., the reactance capacitive). We can in that case tune the array to resonance by inserting an inductance $-jX_A$ at the terminals as shown.

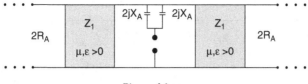

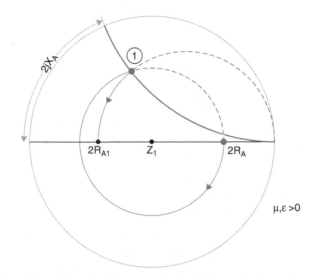

Figure 2.7 By placing dielectric slabs on each side of the array, the radiation resistance $2R_A$ can be transformed into the impedance ①. Further addition in series of the antenna reactance $2jX_A$ lands us in the resistance $2R_{A1}$, yielding a total resistance of R_{A1}. By adjusting Z_1 as well as the slab thickness, we can obtain arbitrary values of R_{A1}.

Alternatively, we may place the two dielectric slabs $\lambda/2$ farther away from the array. Obviously, that leads to the same impedance R_{A1} as obtained in Figure 2.7, but can still lead to the same value as earlier. However, the bandwidth will, in general, be different. If an arbitrary spacing different from $\lambda/2$ is used, matching is still possible, but Z_1 and the slab thickness will be different.

Finally we show in Figure 2.8 a case where the two matching slabs are made of a material with negative μ and ε. In that case the impedance $2R_A$ is rotated counterclockwise in the Smith chart until it reaches ① (see Figure 1.2). From then on, the matching procedure is identical to the case in Figure 2.7 when μ and ε are positive. Whether counterclockwise rotation is actually possible is quite another matter.

From these examples we learn that the terminal impedance $R_A + jX_A$ of an array can be tuned to resonance not only by matching at the terminals,

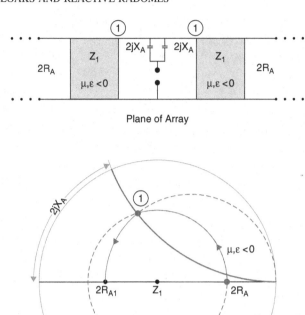

Figure 2.8 The same situation as in Figure 2.7 except that the two slabs have negative μ, ε. Thus, according to conventional theory (see Figure 1.2), the resistance $2R_A$ is rotated counterclockwise to the impedance ①. Addition of $2jX_A$ is similar to the case shown in Figure 2.7.

as usual, but also by placing suitable matching devices such as dielectric slabs on each side of the array. We also saw that the slabs could have μ and $\varepsilon < 0$. This could conceivably lead to a larger bandwidth if designed correctly. However, it should be emphasized that there is, in general, nothing "magical" about using materials with negative μ and ε compared to conventional ones. In both cases the matching can be designed to increase the bandwidth slightly. An example using conventional material is given in Chapter 6 of ref. 3.

2.2.2 Line Arrays and Single Elements

We considered above an infinite planar array flanked by dielectric slabs on both sides. From here we may conceive of many other configurations

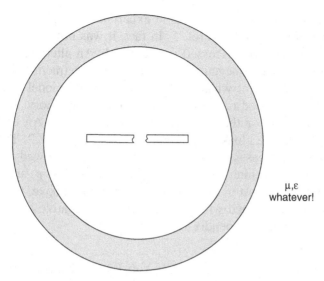

Figure 2.9 Short dipole enclosed in a spherical shell with μ and ε. By variation of the radius of the shell, its thickness, and μ and ε, the dipole can always be matched regardless of the sign of μ and ε. However, the directivity can never exceed (by much!) the directivity of the dipole (see Section 2.3.2).

which work on the same basic principle: use of a dielectric radome that tunes the antenna to resonance. The most obvious is probably a line array surrounded by a dielectric cylindrical shell. Similarly, we may consider a single dipole enclosed in a spherical dielectric shell as illustrated in Figure 2.9. In fact, such a configuration was considered by Ziolkowski [4–6]. However, all his results are simulations based on the existence of negative μ and ε. Many scientists are quite skeptical about these results (see, in particular, Kildahl [7]). Also, Hansen gives a frank assessment in his book about small antennas [8]. Since this author has already commented on materials with μ, $\varepsilon < 0$, we shall leave it at that for now. Again, we emphasize that matching is possible regardless of the sign of μ, ε, but the directivity will basically be as for a dipole as long as the size of the radome is less than $\lambda/4$ (see also Section 2.3.2).

2.3 COMMON MISCONCEPTIONS

2.3.1 Misinterpretation of Calculated Results

When working with metamaterials, it is common practice to assume not only negative μ and ε but also a negative index of refraction. In this way, many interesting but also questionable experiments can be simulated, as amply shown, for example, in ref. 5. In fact, it was how we obtained the concept for the flat lens discussed in Chapter 1. An alternative approach is simply not to assume negative μ, ε, or index of refraction, but simply to calculate scattering for whatever form of the actual model of elements, as would be done based on classical electromagnetic theory, such as the method of moments or the finite element time domain. An example of this approach was the cloak calculation shown in Figure 2.2. This author is aware of several cases where this approach was followed and where the calculations were interpreted as proof of negative μ, ε, and index of refraction. However, just as we saw in the cloak case above, the behavior could always be explained without any notion of negative μ, ε, or index of refraction (see also Appendix A).

2.3.2   Ultimately: What Power Can You Expect from a Short Dipole Encapsulated in a Small Spherical Radome?

Sometimes even experienced antenna practitioners fall into the trap of believing that obtaining maximum voltage from a receiving antenna should be the goal. This is, of course a misconception since merely inserting a transformer could in principle yield any voltage desired. What is important is how much maximum power $\mathbf{P}_{max}$ can be extracted from an antenna when exposed to an incident plane wave with effective field strength E_0 propagating in free space with intrinsic impedance $Z_0 = 120\pi$. The answer is well known:

$$\mathbf{P}_{max} = \frac{E_0^2}{120\pi} A_{eff} = \frac{E_0^2}{120\pi} \frac{\lambda^2}{4\pi} G \qquad (2.1)$$

where A_{eff} is the receiving area of the antenna and G is the gain of the antenna, $G = \varepsilon_A D_A$, where D_A is the directivity of the antenna and ε_A is the efficiency of the antenna. To obtain the maximum power $\mathbf{P}_{max}$, the antenna must be conjugate matched as well as polarization matched.

The antenna efficiency ε_A is determined by the ratio between the loss resistance and the radiation resistance. For a dipole of total length greater than 0.3λ, the efficiency ε_A is almost 100% (if the antenna is made of wire with reasonable good conductivity). For shorter dipoles, the efficiency can typically go down to a few percent, depending on antenna length and wire resistivity. Furthermore, there will (for short dipoles) usually be significant loss in the matching network.

The key to understanding this problem is the directivity D_A. It is obtained by integration of the radiation pattern, which for a short dipole $(2l < 0.1\ \lambda)$ is the well-known doughnut pattern, yielding a directivity $D_A = 1.5$ or 1.76 dBi over an isotropic radiator. There really is nothing you can do to increase D_A except to make the dipole longer, actually quite a bit longer, such as $2l > 0.4\ \lambda$, leading to $D_A \sim 2$ dBi. Thus, when encapsulating a short dipole in any kind of small dielectric shell, it is conceivable (but not likely) that the directivity could increase by 0.1 to 0.2 dBi just because the effective "aperture" has increased. Again, let us emphasize, *you will obtain maximum received power only when conjugate matched*. And therein lies the "rub"! If you compare a short unmatched dipole to the same dipole matched by a dielectric shell or any more conventional matching at the terminals, you can easily obtain fantastic "improvement" in antenna voltage as well as received power. You are simply not comparing apples with apples. It is as misleading as what happened to one of my engineering friends who landed in New York shortly after World War II. Being forced to take whatever job he could find, he ended up working for a small antenna company with somewhat questionable ethical standards. His boss asked him to design a television rabbit ear antenna switch with three positions marked, "Good," "Better," "Best." After a week his boss checked in on him and became quite annoyed when he found that no progress had been made. "Here is what you do," he explained. "In position 'Good' you short circuit the antenna terminals, in position 'Better' you connect just one antenna arm, and in position 'Best' you connect both arms."

It is puzzling that anything as fundamental as a short dipole encapsulated in a dielectric shell has been treated not only in the *IEEE Transactions on Antennas and Propagation*, but also in the magazine, a book section, and probably places that I am not aware of. I do understand the tremendous publication pressure that academicians are coming under today from their department heads and deans: They must publish or perish! Actually, I place more blame on the reviewers than the authors.

2.4   CONCLUDING REMARKS

In a paper written in 1968, Veselago asked a perfectly good question: What would happen if both μ and ε were negative? He examined his question by considering the boundary conditions between an ordinary material with μ, $\varepsilon > 0$ and one with μ, $\varepsilon < 0$. He concluded correctly that it would lead to a negative index of refraction. However, his solution was purely mathematical. The fact is that any time that a mathematical solution is found to a physical problem, it must be tested for physical reality. More specifically, we examined in Chapter 1 perhaps the best known application of a negative index of refraction: the flat lens. We found that to focus inside as well as outside the slab, it should be made of a material that advanced the signal in time: in other words, could produce negative time. Since to the best of this writer's knowledge, no one has ever claimed the existence of such a concept, we state that Veselago's conclusion, although mathematically correct, would not stand up in physical reality.

I am not sure how many scientists are aware of these facts. However, it has not slowed down the steady stream of publications. At least five full-fledged books have been published, as have at least 1000 papers, and this will probably continue for a long time.

Meanwhile, potential contractors are trying to find new applications. The cloaks and radomes discussed in this chapter are but a few examples. Although the usefulness of these gadgets is debatable, one thing is certain: Their functionality does not depend on material with a negative index of refraction. How could it? In Chapter 1 we showed fairly conclusively that such materials simply do not exist!

REFERENCES

[1] M. G. Silverinha, A. Alu, and N. Engheta, Parallel-plate metamaterials for cloaking structures, *Phys. Rev.*, vol. E75, 2007.

[2] B. A. Munk, *Frequency Selective Surfaces: Theory and Design*, Wiley, New York, 2000.

[3] B. A. Munk, *Finite Antenna Arrays and FSS*, Wiley-Interscience, Hoboken, NJ, 2003.

[4] R. W. Ziolkowski and Q. D. Kipple, Application of double negative materials to increase power radiated by electrical small antennas, *IEEE Trans. Antennas Propag.*, vol. 51, Oct. 2003.

[5] N. Engheta and R. W. Ziolkowski, *Metamaterials: Physics and Engineering Explorations*, Wiley-Interscience, Hoboken, NJ, 2006.

[6] A. Alu, N. Engheta, A Erestok, and R. W. Ziolkowski, Single-negative, double-negative and low-index metamaterials and their electromagnetic applications, *IEEE Antennas Propag. Mag.*, vol. 49, pp. 23–36, Feb. 2007.

[7] P-S. Kildahl, Comments on "Application of double negative materials to increase power radiated by electrically small antennas," *IEEE Trans. Antennas Propag.*, vol. 59, p. 766, Feb. 2006.

[8] R. C. Hansen, *Electrically Small, Superdirecting and Superconducting Antennas*, Wiley-Interscience, Hoboken, NJ, 2006, pp. 73–74.

3 Absorbers with Windows

3.1 INTRODUCTION

Many variations of frequency-selective surfaces are known today. Typically, they may reflect at some frequencies while they are transparent at others. Similarly, they can be designed to absorb at some frequencies, whereas they essentially reflect at all other frequencies. In this chapter we consider periodic structures that absorb in certain frequency bands while they are transparent in others. They are often called *rasorbers*. Nobody has yet suggested using metamaterials to solve this problem. So this time we beat them to it!

3.2 STATEMENT OF THE PROBLEM

Absorption can be obtained by placing resistive sheets or elements in front of a ground plane. Such an arrangement can be designed to absorb in a certain frequency range while it will be reflective everywhere else (see Chapter 4). If, in addition, we want our configuration to be transparent in another frequency range, it will be necessary but not sufficient to require that the ground plane be transparent in that range. Basically, this can be done by making the ground plane a slotted FSS. However, we must also require the resistive sheets or elements to be "invisible" at these transparent frequencies while they should absorb otherwise—and therein lies the essential problem. In fact, quite often it is expected that just using some typical lossy FSS elements can somehow do the trick. Although it is sometimes possible to "luck out," it is in general not the way to go. In the following paragraph we show a concept that has been tested and proven to work. However, other approaches may be possible.

Metamaterials: Critique and Alternatives, By Ben A. Munk
Copyright © 2009 John Wiley & Sons, Inc.

3.3 CONCEPT

In Figure 3.1 we show the simplest of all elements: a straight wire exposed to an incident plane wave. If the total element length $2l \sim \lambda/2$, the current distribution will basically be co-sinusoidal and strong, as shown in Figure 3.1b. Similarly, if $2l < \sim \lambda/3$, the current distribution will still be co-sinusoidal but of relatively low amplitude, as shown in Figure 3.1a. Finally, when $2l > 2\lambda/3$, the current distribution will be sinusoidal, complicated (!), and strong, as shown in Figure 3.1c.

When the wire is lossless, only scattering and no absorption will take place. However, if the wire is resistive, the incident field will be partially absorbed and partially scattered. Obviously, when the wire current is negligible as in Figure 3.1a (i.e., the wire segment $2l$ is small compared to $\lambda/2$), neither substantial scattering nor absorption will take place.

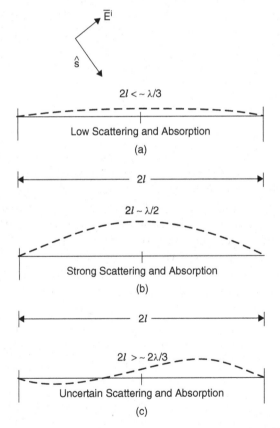

Figure 3.1 Plane wave with direction $\hat{s}$ incident upon a straight wire of length $2l$. Three cases are shown for different wavelengths.

In contrast, substantial absorption and scattering will take place in the cases shown in Figure 3.1b and c.

We may conclude from the presentation above that if a composite element is made from wire segments, we must require that the segments be somewhat shorter than $\lambda/2$ for them to neither scatter nor absorb. In contrast, good absorption (and scattering) will, in general, take place when the wire segments are $2l \sim \lambda/2$. Longer wire segments should be avoided simply because they are somewhat unpredictable and may change dramatically with angle of incidence.

3.4   CONCEPTUAL DESIGNS

Next we show by various examples how to control the current distribution on the elements such that we produce a radome at the middle frequency f_m and an absorber at the lower frequency f_l. A simple case is shown in Figure 3.2, where we have placed a single choke of length $\lambda_m/4$ in the middle of a straight wire. This will effectively prevent any current from flowing across the gaps between the two horizontal sections since a transmission line of length $\lambda_m/4$ short-circuited at one end will exhibit infinite impedance at the other at the midfrequency f_m. Furthermore, we require the two horizontal sections to have length $l < \sim \lambda_m/3$. This will limit the current on these sections to a low value at the frequency f_m (see Figure 3.1a). However, at frequencies below f_m, the input impedance of the choke becomes inductive, whereby the two horizontal sections can be tuned to resonate at a lower frequency f_l, exhibiting a strong current similar to the case shown in Figure 3.1b.

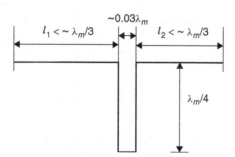

Figure 3.2   When a choke of length $l_m/4$ is inserted midway in a wire of total length 2l, no current can flow across the gap. If $l < \sim\lambda_m/3$, the current on the entire element will be negligible (see Figure 3.1a). λ_m denotes the wavelength at the midfrequency f_m. However, at a lower frequency f_l, the choke becomes inductive and a strong current can exist on the entire element (see Figure 3.1b).

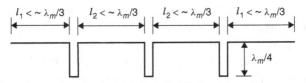

Figure 3.3 When more chokes are added as shown, a strong current over the entire element can be obtained at a frequency f_l that is lower than for the one-choke case shown in Figure 3.2.

If we use only a single stub, as shown in Figure 3.2, it is obvious that the length requirements of the horizontal sections can only produce a lower resonant frequency f_l, somewhat below the middle frequency f_m. However, to lower f_l further below f_m, we can use more chokes and more horizontal sections, as shown in Figure 3.3 for three chokes. We must again require all the chokes to have the same length, $\lambda_m/4$. However, the horizontal sections l_1 and l_2 do not necessarily have to be equal, but they must all be smaller than about $\lambda_m/3$.

Minor variations of the attachments of the chokes to the horizontal wires are possible. For example, the horizontal sections can be attached to the chokes anywhere, as shown in Figure 3.4, as long as the total length of the chokes is $\lambda_m/4$. When connected to the open end, the chokes exhibit the largest bandwidth, whereas it is zero when connected to the short-circuited end. Another motivation for alternative attachment points may simply be related to space limitations (see, e.g., Figure 3.7).

We have not shown the current distribution at the middle frequency f_m simply because it is small (ideally, it is zero, or at least the average is). However, at the low frequency f_l the current is very strong, as shown in Figure 3.5 for the three-choke case. This, of course, is exactly what we need to obtain an effective absorption. In fact, we merely have to add some resistance to the elements. This can be done in two basic ways: We can make the horizontal sections out of a resistive material (the chokes

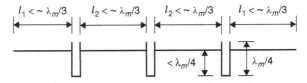

Figure 3.4 The bandwidth of the chokes can be varied by variation of the attachment point of the horizontal sections. The closer the point is to the open end of the chokes, the more bandwidth there is when choking. The total length of the chokes must always be $\lambda_m/4$.

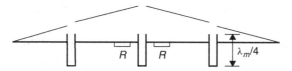

Figure 3.5 Typical current distribution I_l on a three-choke design at the low frequency f_l. The load resistors R are placed close to the middle choke, where the current is maximum at the frequency f_l and negligible at f_m.

must always be made of a perfect conductor to work effectively), or we can insert lumped resistors, as denoted by R in Figure 3.5. The latter is the preferred approach because these resistors can be placed where the current at the frequency f_l is maximum (i.e., right next to the choke in the middle). This position has the added advantage that the current at the frequency f_m is closest to zero there, such that minimum attenuation is encountered in the transparent band.

We finally place the choked FSS in front of a slotted ground plane that is transparent at f_m, as shown in Figure 3.6. Typically, the spacing between these two components should be $\lambda_l/4$. That leads to a spacing that is considerable at the frequency f_m. However, since both the ground plane and the choked FSS are invisible (at least in principle), this distance is of only minor importance.

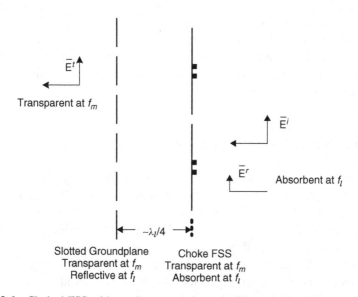

Figure 3.6 Choked FSS with an element as shown in Figure 3.5 is placed a distance of about $\lambda_l/4$ in front of a ground plane that is transparent at the midfrequency f_m.

3.5   EXTENSION TO ARBITRARY POLARIZATION

So far we have considered only designs capable of handling linear polarization. In this section we show how to modify the structure such that arbitrary polarization can be accommodated. It is often believed that it can be done simply by placing two linear systems as above, orthogonal to each other. Although at first this sounds like a good idea, the reader should be warned that this approach may encounter certain practical limitations. The fundamental problem is typically how to place the elements close enough to each other that grating lobes can be avoided at the midfrequency f_m. Since these can be of high intensity and at some frequencies have directions coinciding with the incident field, they can severely jeopardize the backscattered field and must therefore be avoided at all costs. Further, since the onset of grating lobes for a specific angle of incidence and array configuration depends only on the interelement spacings, it becomes of paramount importance to choose an array type that will enable us to delay the onset of grating lobes as much as possible.

One of the most efficient ways to pack elements close together is to use three-legged elements arrayed in a triangular grid. Thus, based on what we have seen above, we choose the choked FSS design shown in Figure 3.7. We observe that each of the element legs has two chokes. To reduce direct coupling between neighboring chokes, the attachment points have been moved slightly down from the open end of the chokes as discussed earlier. Similarly, the end of the legs have been shaped like "anchors" in order to make the elements as compact as possible (i.e., the onset of grating lobes is delayed). Finally, we have added load resistors in each leg as shown. The dimensions given in Figure 3.7 are approximately such that absorption should take place in the X-band and transparency should be in the K_A-band.

3.6   THE HIGH-FREQUENCY BAND

So far not much has been said about the high-frequency band f_h except that it is complicated and to some extent unpredictable. We are, of course, talking here about the absorption and scattering properties of the large composite rasorber element, as depicted, for example, in Figure 3.7. However, it really is not terribly important provided that we use the right design approach. In fact, we merely place an FSS resonating at f_h and loaded with resistors (or simply made of resistive material) in front of the entire configuration. This is backed by another FSS, also resonating at f_h but with no resistive loading (i.e., it acts as a ground plane at f_h). The

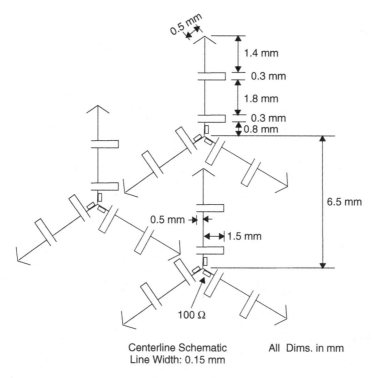

Figure 3.7   Example showing how the elements in Figure 3.5 can be modified to handle arbitrary polarization. Three-legged elements arranged in a triangular grid are particularly well suited since the elements can be closely packed and thereby delay the onset of grating lobes.

electrical spacing between these two FSSs should typically be around $\lambda_h/4$ to provide good absorption at f_h. However, at the lower frequencies, f_l and f_m, these two FSSs are basically invisible (actually, they are slightly capacitive, which can be tuned out). Note that these two FSSs act as a shield at f_h such that no grating lobes from the rasorber panel will be excited in that band.

3.7   COMPLETE CONCEPTUAL RASORBER DESIGN

A complete conceptual rasorber design is shown in Figure 3.8. The signal is incident from the left. We first encounter FSS panels ① and ② both resonating in the high-frequency band f_h. Panel ① is loaded with resistors, while ② acts as a ground plane (i.e., together they act as an absorber in the high band, as explained above). Note that these two panels are essentially

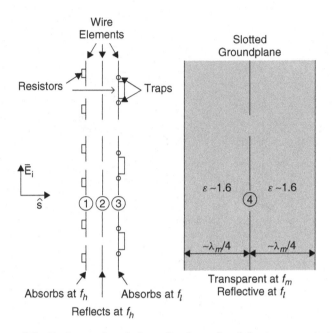

Figure 3.8   Basic rasorber design: absorbs at f_l and f_h; transparent at f_m.

transparent at the lower frequencies f_l and f_m. Next, we see rasorber panel ③ as shown in Figure 3.7; that is, it will absorb in the low-frequency band f_l in conjunction with the slotted ground plane ④, while both of these panels are transparent at the midfrequency band f_m. We note finally that the slotted screen ④ is flanked by a dielectric slab on each side. Their thickness is slightly larger than $0.25\lambda_m$ electrically and their purpose is to provide scan compensation at the radome frequency f_m. The value of the dielectric constant should typically be around 1.6, as explained in ref. 1.

Note further that the rasorber panel ③ must be supported by a dielectric substrate that will affect the transmission as well as the backscatter. At the midfrequency f_m it is suggested that the capacitive effect of the substrate be canceled, simply by making the traps slightly more inductive: for example, by making them slightly shorter. It is strongly recommended first to tune rasorber panel ③ alone to perfect transmission.

At the low frequency f_l, both the rasorber substrate and the dielectric slab in front of the slotted ground plane will affect the optimum spacing between the rasorber panel ③ and the slotted ground plane ④. The simplest way to overcome this problem is, in general, to reduce the spacing from about $0.25\lambda_l$ to something like $0.20\lambda_l$. From a practical point of view, substrates could be placed between sheets ① and ② as well as between sheets ② and ③. If $f_h \sim 100$ GHz, typical thicknesses for each layer could be around 20 mils.

3.8 PRACTICAL DESIGNS

The concepts and design approaches outlined above have been used to calculate several designs. They were all very successful and worked well as both absorbers and radomes. Unfortunately, permission to show them was not given. Thus, we merely show in Figure 3.9 the typical performance of a panel without a frequency or dimensions.

3.9 OTHER APPLICATIONS OF TRAPS: MULTIBAND ARRAYS

We saw above how traps can be used to produce panels that are transparent at some frequencies and absorptive at others. There are, however, other applications. For example, it is sometimes desirable to design an interlaced dual-band array leading to short elements operating at the high-frequency band and long elements at the low band. While the short elements, in general, will not "bother" the long elements operating in the low-frequency band (see Figure 3.1a), the opposite situation can sometimes be problematic. One solution to this problem is simply to "kill" the current on the long elements at high frequencies by use of traps, as illustrated, for example, in Figures 3.3 and 3.4.

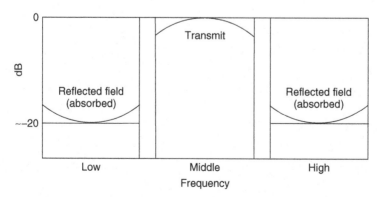

Figure 3.9 Typical reflection in the low and high bands where absorption takes place. Also, typical transmission at midband where the panel works like a radome.

The concept can eventually be extended to multiband arrays. However, the impedance properties of all the traps must be evaluated carefully in all frequency bands.

REFERENCE

[1] B. A. Munk, *Frequency Selective Surfaces: Theory and Design*, Wiley, New York, 2000, Chaps. 4 and 5.

4 On Designing Absorbers for an Oblique Angle of Incidence*

4.1 LAGARKOV'S AND CLASSICAL DESIGNS

At the International Conference on Materials for Advanced Technologies (ICMAT 2007) in Singapore, the well-known Russian professor A. Lagarkov and his co-worker V. Kisel presented a new and intriguing absorber concept based on metamaterial with a negative index of refraction. Their claim was that their design should be able to absorb all incident signals, regardless of angle of incidence and frequency.

They introduced their concept as illustrated in Figure 4.1. Here a line source is placed at a distance, y_0, from an impedance wall (i.e., an absorber). They then ask: How should this impedance be designed to ensure transfer of maximum energy into the absorber, or by implication, how can we reduce the signals reflected from this interface as much as possible for all angles of incidence? The actual Lagarkov concept is shown in Figure 4.2. It consists basically of a thin resistive sheet presumably with a sheet resistance equal to 120π (space cloth). It is backed by an airspace followed by a flat lens, which again is followed by a ground plane as shown in the figure. The airspace and the flat lens have the same thickness. Thus, all incident signals will, according to the generally accepted theory discussed in Section 1.2, be delayed as much in the airspace as it is advanced in the flat lens, regardless of the angle of incidence or frequency. (Indeed, some of us have a hard time accepting this!) In other words, all signals reflected from the ground plane will meet at the "focal point" with the same phase delay, π, as caused by the reflection from the ground plane.

*This chapter is based partly on a paper published in *IEEE Transactions on Antennas and Propagation* in January 2007 coauthored by B. Munk, P. Munk, and J. Pryor.

Metamaterials: Critique and Alternatives, By Ben A. Munk
Copyright © 2009 John Wiley & Sons, Inc.

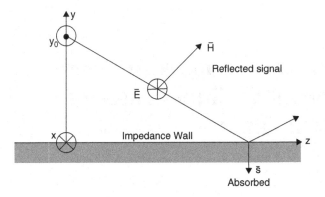

Figure 4.1 The way that Lagarkov analyzes his absorber: a line current at y_0 radiating down into an impedance wall.

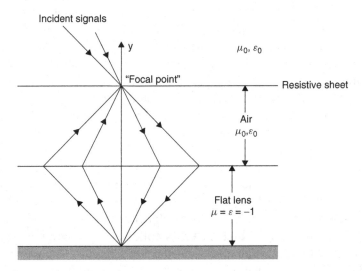

Figure 4.2 The concept behind Lagarkov's absorber design: The flat lens advances the phase as much as the airspace delays it. Thus, all signals reflected from the ground plane will be 180° out of phase at all angles of incidence and all frequencies at the focal point.

We may characterize this design as a Salisbury screen, where the effect of the ground plane has been eliminated by use of a flat lens. Later we investigate ordinary Salisbury screens and find that they are indeed affected severely by the ground plane, leading to limited bandwidth and angle of incidence dependence. Thus, if a flat lens could be realized, there is no question that we would indeed be witnessing a major breakthrough.

However, there are at least two serious flaws in Lagarkov's design. First, nobody has yet been able to actually build a working flat lens, as far as we know. It exists only as a concept in some people's minds. In fact, this author categorically rejects its existence, as discussed in Sections 1.10 and 1.11 as well as in Chapter 6. As stated there, it will simply require "negative" time, which is okay from a mathematical, but not from a physical, point of view. Of course, we can use amplifiers with delay times. That leads to phased arrays, not to new materials.

But even if a flat lens could be realized, we would not yet be out of the woods! More precisely, if the space cloth has a sheet impedance of 120π, that would produce zero reflection only for a normal angle of incidence. For oblique incidence, the sheet impedance should be lowered by the cosine to the angle of incidence for parallel polarization, and similarly, be increased by the inverse for orthogonal polarization. This is all discussed in great detail in the following sections, where we also show what to do about this problem: namely, use of dielectric matching plates. Typically, they have a dielectric constant in the range 1.3 to 1.8 (i.e., they are fully realizable, in contrast to materials with $\mu = \varepsilon = -1$).

In the next sections we investigate various absorber designs, such as the Salisbury screen, the Jaumman absorber, and the circuit analog absorber. They have all been designed to yield good absorption for all angles of incidence, ranging from normal up to $\pm 45°$ for both orthogonal and parallel polarization. Like Lagarkov, we use a matching plate, but ours has $\varepsilon = 1.6$ and $\mu = 1$. Thus, ours is indeed more readily realizable and we present real calculated results showing a typical bandwidth of an octave for 20 dB or better absorption for both orthogonal and parallel polarization.

Absorbers can be constructed by placing one or more resistive sheets in a stratified medium. The simplest design, called a *Salisbury screen* after its inventor [1], consists of a single resistive sheet placed $\lambda/4$ in front of a perfectly conducting ground plane. Extension of the bandwidth is possible by use of more resistive sheets spaced approximately $\lambda/4$ apart. These types are usually referred to as *Jaumann absorbers* [2, Sec. 9.3]. Further increase in bandwidth is possible by placing suitable dielectric slabs between the resistive sheets and, in particular, in front of the outermost resistive sheet. Finally, the resistive sheets can be made in the form of lossy FSS sheets. That will result in a complex sheet admittance that by proper design can extend the bandwidth even further [2, Sec. 9.4].

Examples of all these types of absorbers exposed to an incident field with arbitrary polarization for normal as well as oblique incidence are given below. Note that here we are interested primarily in specular reflection. For bistatic scattering, in general, see Chapters 4 and 5 of ref. 3.

4.2 SALISBURY SCREEN

The Salisbury screen, shown in the insert of Figure 4.3a, will serve as our baseline design. It consists of a single resistive sheet with conductivity Y_S equal to that of free space Y_0 and is placed a distance d_1 in front of a perfectly conducting ground plane. The relative dielectric constant of slab d_1 is denoted ε_1. Also, in Figure 4.3a we show the specular reflection

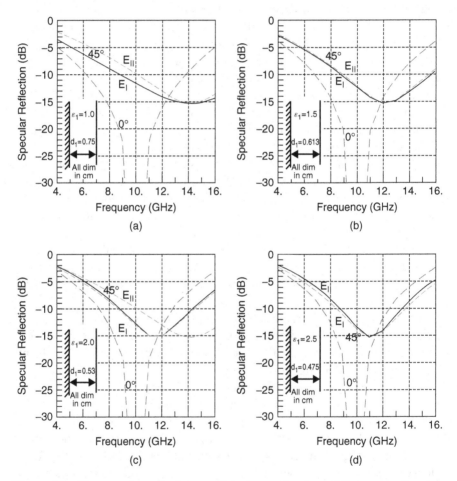

Figure 4.3 The specular reflection coefficient as a function of frequency of various Salisbury screens for normal and 45° angles of incidence. The dielectric constant of the medium between the ground plane and the resistive sheet is as shown in the respective inserts: (a) $\varepsilon_1 = 1.0$ with $d_1 = 0.75$ cm; (b) $\varepsilon_1 = 1.5$ with $d_1 = 0.613$ cm; (c) $\varepsilon_1 = 2.0$ with $d_1 = 0.53$ cm; and (d) $\varepsilon_1 = 2.5$ with $d_1 = 0.475$ cm. Note: At 10 GHz, $\beta_1 d_1 = \pi/2$ for all cases. Further, the resistive sheet has conductance $Y_s = Y_0$.

curves for a normal as well as a $45°$ angle of incidence (orthogonal and parallel polarization). We observe immediately two fundamental dilemmas for an oblique angle of incidence:

1. The resonant frequency is raised from 10 GHz to about 14 GHz.
2. The specular reflection coefficient changes from $-\infty$ dB to -15 dB.

The change in resonant frequency is easy to explain. Let the space d_1 have the propagation constant β_1, while r_{1y} denotes the direction cosine* of the wave propagation inside the dielectric. Then the electrical spacing is equal to $\beta_1 d_1 r_{1y}$ [2, Secs. 4.6, 4.7, and 7.3 and Chap. 5], and resonance occurs when $\beta_1 d_1 r_{1y} = \pi/2$. When the angle of incidence is changed from normal to some oblique value, r_{1y} will change from 1 to a value smaller than 1, depending on ε_1. For air, $r_{1y} = 0.707$ at a $45°$ angle of incidence (i.e., resonance occurs at $10/0.707 = 14.14$ GHz), as observed in Figure 4.3a. If we fill the space d_1 with a dielectric medium, r_{1y} will for the same angle of incidence in air be reduced as a direct consequence of Snell's law. Thus, the change in resonant frequency will also be reduced. In fact, we show in Figure 4.3a–d a series of reflection curves where we have varied the relative dielectric constant ε_1 in steps from 1.0 to 2.5 and where the electrical thickness $\beta_1 d_1 = \pi/2$ at $f = 10$ GHz for all values of ε_1. Clearly, we observe a significant stabilization of the resonant frequency with angle of incidence as we increase ε_1. However, we also observe some reduction in bandwidth as ε_1 increases. This is simply because the intrinsic admittance $Y_1 = \sqrt{\varepsilon_1/\mu_1}$ increases with ε_1. Thus, ε_1 should not be chosen arbitrarily high. Values around 2 or slightly higher seem a good compromise for many practical designs, as will be shown later by examples.

We also observe in Figure 4.3a–d that the resonance reflection coefficient at a $45°$ angle of incidence remains at ~ -15 dB regardless of ε_1 and polarization. The reason for this is quite simple. At resonance, $\beta_1 d_1 r_{1y} = \pi/2$; that is, the ground plane admittance ①, as shown later in Figure 4.4, is zero, such that the incident field sees only the sheet admittance, $Y_S = Y_0$. Thus, the bistatic reflection coefficient for orthogonal and parallel polarization, respectively, with respect to the E-field is then [2, App. C]

---

*The direction cosine r_{1y} is defined as the cosine to the angle between the direction(s) of propagation $\hat{r}$ and a specific axis, in this case the y-axis. The subscript 1 refers to medium 1.

$$\overset{E}{\underset{\perp}{\Gamma}} = \frac{Z_0 - Z_0/r_{0y}}{Z_0 + Z_0/r_{0y}} \tag{4.1}$$

$$\overset{E}{\underset{\parallel}{\Gamma}} = \frac{Z_0 - Z_0 r_{0y}}{Z_0 + Z_0 r_{0y}} \tag{4.2}$$

For a 45° angle of incidence we have $r_{0y} = 0.707$ and we obtain $\overset{E}{\underset{\perp}{\Gamma}} = -\overset{E}{\underset{\parallel}{\Gamma}} = -0.172$ or -15.3 dB below the ground-plane reflection, which is precisely what is observed in Figure 4.3b–d for all values of the dielectric constant ε_1 and both polarizations.

4.3   SCAN COMPENSATION

In Section 4.2 we observed a reflection at oblique incidence caused by a mismatch between free space and the sheet admittance. If the match were perfect at a normal angle of incidence, the reflection would increase significantly with scan angle. This phenomenon is by no means unique for planar absorbers. Actually, we observe exactly the same phenomenon when a plane wave is incident upon an array of dipoles: for example, is loaded with resistors (see, e.g., ref. 2, Sec. 9.9.2, 4.11.2, and 5.10). Furthermore, we show at the same place that the scan impedance could be made approximately constant by placing a suitable dielectric slab in front of the dipoles.

It is worth pointing out that this compensation is not a simple consequence of Snell's law, as is often assumed. In fact, for oblique incidence the electrical thickness should be slightly thicker than one-fourth wavelength, while the optimum dielectric constant is given approximately by [2, Sec. 5.10, eq. (5.52)]

$$\varepsilon_1 \approx 1 + \cos \theta_0 = 1 + r_{0y} \tag{4.3}$$

where θ_0 is the angle between the direction of propagation in air and the normal to the array.

Thus, guided by our experience from arrays, we place a dielectric slab in front of the resistive sheet. Since the optimum angle of incidence is chosen here to be 45°, we obtain, according to (4.3), the dielectric constant $\varepsilon_1 \sim 1 + \cos 45° = 1.7$ and the electrical thickness is chosen to be about $0.25\lambda_2$ (i.e., the mechanical thickness at 10 GHz is equal to 0.574 cm, as shown in the insert of Figure 4.4).

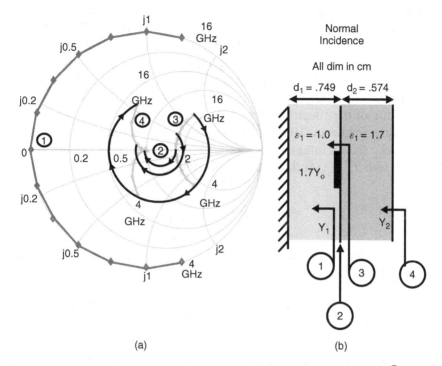

(a)   (b)

Figure 4.4 (a) Smith chart depicting various admittances as shown in (b): ① denotes the ground-plane admittance as seen at the plane of the resistive sheet looking toward the ground plane; ② is the admittance of the resistive sheet; ③ is the sum of ① and ②; ④ is obtained by transforming ③ through slab d_2. (b) Schematic of a Salisbury screen with dielectric slab d_2 in front. Note: Frequency compensation is obtained because the various frequencies of ③ are rotated into ④ such that the wideband frequencies cluster around the center of the Smith chart. Note: The circled numbers denote the admittances at the various positions in the absorbers, not the numbers of layers.

4.4   FREQUENCY COMPENSATION

In Section 4.3 we suggested placing a dielectric matching plate in front of the resistive sheet. Our initial motivation was to obtain scan compensation. However, in this section we show that it is also possible by proper design to achieve a significant increase in bandwidth (i.e., we can obtain frequency as well as scan compensation at the same time; for more details, see Chapter 6 of ref. 3).

The equivalent circuit for such an arrangement is shown in Figure 4.4, where we also show a Smith chart with the relevant admittances as follows. The admittance as seen looking toward the ground plane at the plane of the resistive sheet is denoted ①. It is purely imaginary and therefore located at the rim of the Smith chart. Further, the sheet admittance Y_S is denoted ② and is seen no longer equal to Y_0 (its optimum value will be determined below). The sum of ① and ② is denoted ③, which is finally transformed through slab d_2 into the admittance curve ④.

The value of the sheet admittance ② is obtained simply by noting that at the center frequency 10 GHz, the ground-plane admittance ① is equal to zero (i.e., the admittance sum ③ is simply equal to ②). Thus, if the dielectric slab d_2 is $\lambda_2/4$ thick, the input admittance ④ is given simply by

$$\text{curve } ④ = \frac{Y_2^2}{\text{curve } ②} \tag{4.4}$$

For a perfect match we must require curve ④ $= Y_0$ at resonance, and from (4.4) we obtain, for $\varepsilon_2 = 1.7$,

$$\text{curve } ② = \frac{Y_2^2}{Y_0} = 1.7Y_0 \tag{4.5}$$

Also shown in the Smith chart of Figure 4.4 are the transformation circles for the specific frequencies 10, 13, and 16 GHz.* Note that the admittance at the center frequency, 10 GHz, is rotated exactly $\lambda_2/4$, while the higher frequencies, 13 and 16 GHz, are rotated more and more as the frequency increases (similarly, the lower frequencies will rotate less and less). As seen in Figure 4.4, this results not only in a transformation of the admittance curve ③ into ④ in the center of the Smith chart but also in a compression around Y_0 at the center frequencies (i.e., we obtain a larger bandwidth). This is a typical example of broadband matching. For more about this subject, see Chapter 6 of ref. 3.

Further, we show in Figure 4.5 the reflection coefficient as a function of frequency for normal and 45°. At a normal angle of incidence we observe a substantial increase in bandwidth compared with the simple Salisbury screen shown in Figure 4.3a; in fact, the 20-dB bandwidth has been increased by a factor of about 2.3. Further, we show in Figure 4.6 the reflection curves at a 45° angle of incidence for orthogonal as well as parallel polarization. They are seen to be lacking somewhat in both frequency stability and bandwidth, in particular for the parallel polarization. However, as demonstrated in Figure 4.3, this problem can be partly cured

---

*For further discussion of this admittedly somewhat unorthodox approach, see Appendix B of ref. 3.

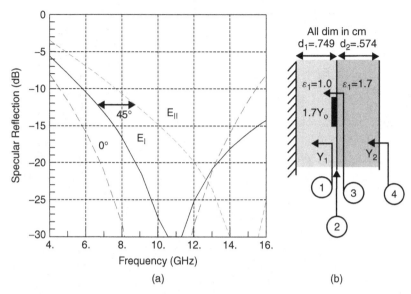

(a)   (b)

Figure 4.5 (a) Specular reflection coefficient as a function of frequency for normal and 45° angles of incidence, orthogonal as well as parallel polarization. (b) Schematic of a Salisbury screen with a dielectric slab d_2 in front. Note the significant shift of resonant frequency for oblique incidence. It is due primarily to $\varepsilon_1 = 1$, as illustrated in Figure 4.3. This problem is alleviated by an increase in ε_1, as shown in Figure 4.6.

by increasing ε_1. Further, the length of d_1 and d_2 should be made slightly longer than $\lambda_1/4$ and $\lambda_2/4$, respectively. This will partly compensate for the reduction caused by r_{1y} and r_{2y} with angle of incidence as discussed earlier.

Thus, based on these considerations, we arrive at the design shown in Figure 4.6. We observe a much improved frequency stability with angle of incidence compared with the earlier design shown in Figure 4.5, although the bandwidth for parallel polarization is still somewhat lacking.

Probably the best guidance of design is to observe the various admittances in the complex plane: in particular, the Smith chart, shown in Figure 4.7a for a normal angle of incidence. Note that at 10 GHz the admittance ④ has been moved slightly to the right by choosing $Y_S = 1.6Y_0$ and not $1.7Y_0$, as calculated originally in equation (4.5). That simply centers the other frequencies of ④ a little better (i.e., we obtain a lower reflection for more frequencies). Finally, the oblique cases for 45° are shown in Figure 4.7b and c for orthogonal and parallel polarization, respectively. Note in the orthogonal case in Figure 4.7b how ② is increased by $1/\cos 45°$ to $1.7 \cdot \sqrt{2} = 2.4$. However, the intrinsic admittance of slab d_2

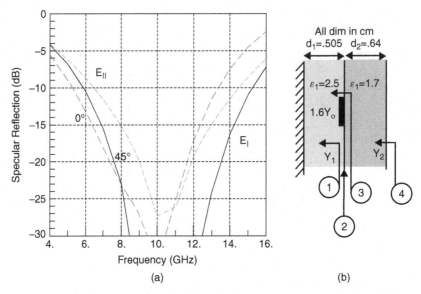

Figure 4.6 (a) Specular reflection coefficient as a function of frequency for normal and 45° angles of incidence, orthogonal as well as parallel polarization. (b) Schematic of a Salisbury screen with a dielectric slab $d_2 = 0.64$ cm in front and $d_1 = 0.505$ cm behind. This design shows much better stability with angle of incidence than the design shown in Figure 4.5 because $\varepsilon_1 = 2.5$ instead of $\varepsilon_1 = 1.0$.

is also increased by 1/cos 45° such that ④ still lands around Y_0. Similarly, we observe in Figure 4.5c for the parallel case how ② is reduced by cos 45° and so is the intrinsic admittance for slab d_2, such that ④ again lands around Y_0, as in the orthogonal case. Alternatively, we can simply prefer not to renormalize the admittance by 1/cos 45° and cos 45°, respectively. This approach is followed in the next section, where we consider the circuit analog absorber.

4.5 CIRCUIT ANALOG ABSORBERS

The Jaumann absorbers discussed above consisted of simple resistive sheets where the sheet admittance ② was independent of frequency. In a circuit analog (CA) absorber these sheets are basically substituted for by resistive sheets with a frequency-selective pattern. Such a combination is capable of producing a complex sheet admittance ② that can greatly enhance the bandwidth beyond what is possible with a simple Jaumann absorber.

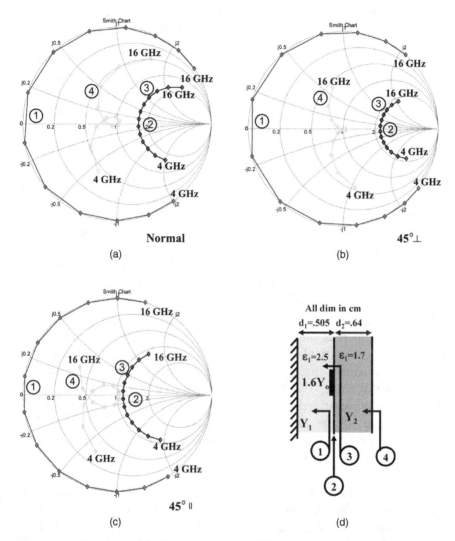

Figure 4.7 Various Smith chart plots for the Salisbury screen shown in the insert (same as in Figure 4.6). ① denotes the ground-plane admittance as seen at the plane of the resistive sheet looking toward the ground plane; ② is the admittance of the resistive sheet; ③ denotes the sum of ① and ②; and ④ is obtained by transforming ③ through slab d_2. Note the frequency compensation taking place by transformation through d_2. (a) Smith chart plot for normal incidence; (b) Smith chart plot for 45° incidence, orthogonal polarization; (c) Smith chart plot for the 45° incidence, parallel polarization.

Basically, any classical FSS element is a candidate for a CA sheet; in fact, a rather complete survey and discussion of elements in general can be found in Chapter 2 of ref. 2. Traditionally, most CA absorbers have been made of straight dipole elements. Since absorbers in general are required to work for arbitrary polarization, the elements typically were made of crossed or orthogonal clusters of straight elements, as explained in Section 9.5.4 of ref. 2. Such configurations are capable of producing absorbers with a bandwidth in excess of 10:1. However, these designs are intended to produce a low return in the backscatter region for normal and moderate angles of incidence only. In the present case we require the bistatic return, as well, to be low for angles ranging from normal up to a ±45° angle of incidence. To avoid the onset of free-space grating lobes (which can severely jeopardize the bistatic return) and to some extent even trapped grating lobes, we must simply require an interelement spacing smaller than $\lambda_0/2$, where λ_0 is the free-space wavelength at the highest frequency. It is possible to embed a dipole array in thin dielectric slabs ("underwear") with high ε and thereby delay the onset of free-space grating lobes [2, Sec. 5.13.1]. However, trapped grating lobes easily occur for an oblique angle of incidence, making precise matching difficult. In general, it is a better idea to use elements that are inherently small. As emphasized numerous times in Sections 2.3 and 2.7 of ref. 2, elements of the loop type are particularly suitable because of their small size, leading to small interelement spacings. Here we use a square loop, where the side length is only about $\lambda_0/4$ at the center frequency (it will be reduced further when placed in a stratified medium, as shown later). More specifically, we show in Figure 4.8 the dimensions of an array of square loops as well as the dielectric profile. Also shown is the reflection coefficient as a function of frequency for a normal incidence. For square loop designs, see also ref. 4.

A more detailed explanation of the intricacies of this CA absorber is obtained by plotting the appropriate curves in a Smith chart as obtained from the PMM program. Thus, we show in Figure 4.9a the ground-plane admittance ① as well as the CA sheet admittance ② for a normal angle of incidence. Note that the latter is no longer constant, as was the case with the Jaumann absorber, but varies strongly with frequency. In fact, when the ground-plane admittance ① is inductive below about 10 GHz, the CA admittance is capacitive. Thus, when these two curves are added together to yield the total admittance ③, as shown in Figure 4.9b, we notice that this curve is somewhat compressed as a function of frequency because the reactive parts of ① and ② partially cancel each other.

Finally, we transform ③ through slab d_2 and obtain the final curve ④, as also depicted in Figure 4.9b. We note again further compression simply

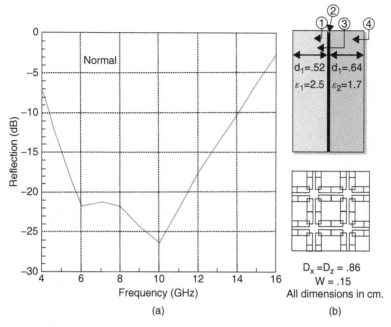

Figure 4.8   The reflection coefficient as a function of frequency for the CA design shown in part (b), normal angle of incidence, as obtained using the PMM program. The CA sheet is comprised of perfectly conducting square loops loaded with eight 100-Ω resistors, one in each corner and one in the middle of each side.

because the higher frequencies rotate faster than the lower ones, just as was the case with the Jaumann absorber discussed earlier. Similarly, we show in Figure 4.10 the reflection coefficient curves for a 45° angle of incidence at orthogonal as well as parallel polarization. Further, curves ① to ④ are shown in two Smith charts in Figure 4.11a and b for orthogonal polarization. And finally, we show the same curves in Figure 4.12a and b for parallel polarization. The compression of the final curve ④ is seen to be even better than for a normal angle of incidence simply because the oblique case has been favored in the design procedure. Note how the frequency point 16 GHz of curve ② in Figure 4.12a apparently exhibits a discontinuity. However, the same curve but with a finer frequency increment is shown in Figure 4.13 and shows total continuity. A closer investigation shows that this anomaly is caused by excitation of the odd mode on the loops at about 20 GHz. It produces an antiresonance approximately midway between the fundamental resonance at 10 GHz and the second resonance at about 20 GHz (see also ref. 2, Chap. 2).

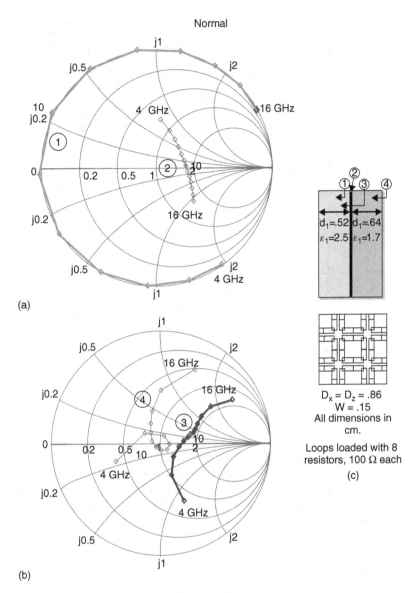

Figure 4.9   Smith chart plots of the CA design shown in part (c) for the normal angle of incidence as obtained with the PMM code. (a) Admittance ① denotes the ground-plane admittance as seen in the CA sheet looking left. Admittance ② denotes the admittance of the CA sheet alone; it is comprised of an array of conducting square loops loaded with eight 100-Ω resistors. (b) Admittance ③ denotes the sum of ① and ② shown in Figure 4.9a. Admittance ④ denotes the final input admittance as seen at the surface of slab d_2 looking left; it is obtained by transforming ③ through slab d_2.

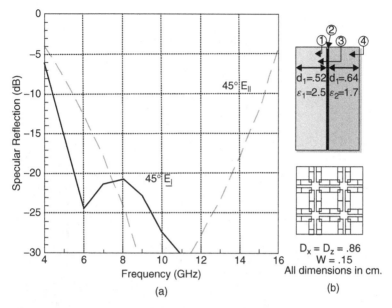

(a)   (b)

Figure 4.10   Specular reflection coefficient as a function of frequency for the CA design shown in part (b) for a 45° angle of incidence, orthogonal ($\perp$) as well as parallel ($\parallel$) polarization as obtained with the PMM code. The CA sheet is comprised of perfectly conducting square loops loaded with eight 100-Ω resistors, one in each corner and one in the middle of each side.

4.6   OTHER DESIGNS: COMPARISON AND DISCUSSION

Comparison with other designs is always of interest, in particular if a different approach has been used. Particularly noteworthy in that respect are two papers that use the microgenetic algorithm (MGA) approach. The first of these [5] appeared in June 2001 and with the exception of a single design contains only examples that are optimized for a single polarization. However, the second paper [6], which appeared in March 2002, contains designs that are optimized for arbitrary polarization, as considered exclusively in this chapter. Consequently, only two examples from the latest paper are considered here, denoted cases 1 and 2.

Both cases contains two FSS screens based on the "pixel" approach. They are sandwiched between several dielectric slabs, some lossy and some essentially lossless. Both designs have about a 19-dB bandwidth, slightly less than an octave. The lowest bistatic reflection coefficient is -19.38 and -18.45 dB for cases 1 and 2, respectively. Further, the total thickness is 5.7 and 5.65 mm, respectively, corresponding to $0.57\lambda_0$ and $0.565\lambda_0$ at 30 GHz. These thicknesses are quite comparable to all the

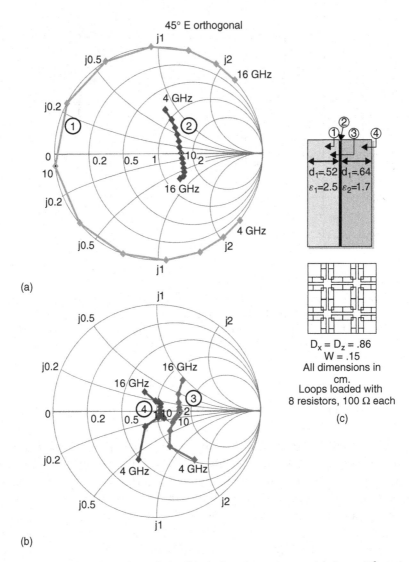

Figure 4.11 Smith chart plots of the CA design shown in part (c) for a 45° angle of incidence, orthogonal E-field polarization. (a) Admittance ① denotes the ground-plane admittance as seen at the CA sheet looking left. Admittance ② denotes the admittance of the CA sheet alone; it is comprised of an array of conducting squared loops loaded with eight 100-Ω resistors. (b) Admittance ③ denotes the sum of ① and ② shown in Figure 4.11a. Admittance ④ denotes the final input admittance as seen at the surface of slab d_2 looking left; it is obtained by transforming ③ through slab d_2.

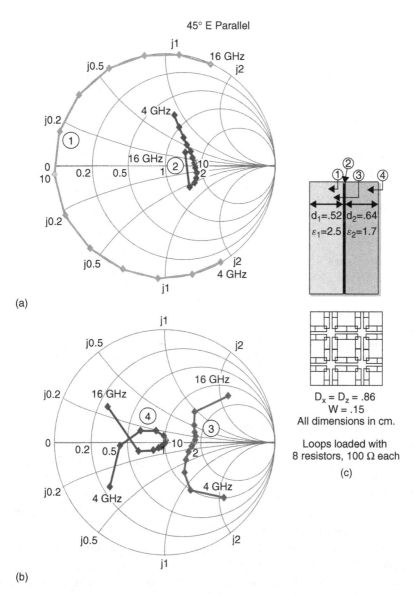

Figure 4.12 Smith chart plots of the CA design shown in part (c) for a 45° angle of incidence, parallel E-field polarization as obtained with the PMM code. (a) Admittance ① denotes the ground-plane admittance as seen at the CA sheet looking left. Admittance ② denotes the admittance of the CA sheet alone; it is comprised of an array of conducting squared loops loaded with eight 100-Ω resistors. (b) Admittance ③ denotes the sum of ① and ② shown in Figure 4.12a. Admittance ④ denotes the final input admittance as seen at the surface of slab d_2 looking left; it is obtained by transforming ③ through slab d_2.

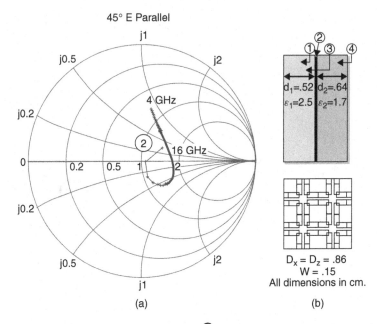

(a)   (b)

Figure 4.13 The same CA admittance curve ② as shown in Figure 4.12a but with finer frequency intervals. It clearly shows continuity. It is related to an antiresonance between the fundamental resonance at 10 GHz and the second resonance around 20 GHz (odd mode).

designs presented in this chapter. The present design has a bandwidth about the same as that in MGA designs; however, the reflection is significantly lower. For orthogonal polarization the bandwidth is significantly larger ($f_H/f_L \sim 2.7$).

The difference in performance between the design obtained here by the analytic approach and that in the MGA approach, obtained by more than 31 hours of computer optimization per case, is interesting to explore. First, the computer placed FSS screen 2 only about $0.03\lambda_0$ in front of the perfectly conducting ground plane for case 2. This spacing is simply so small that it essentially shorts out FSS screen 2 and thus renders it ineffective. In fact, if FSS screen 2 next to the ground plane is removed completely, results obtained from the PMM program change by less than 1 dB. Similarly, for case 1 the analog spacing was somewhat longer but still smaller than desirable to be effective. FSS screen 1, on the other hand, is placed in the "right neighborhood" to be effective provided that it is designed correctly. However, closer inspection shows that the surface resistances of both CA screens are about an order of magnitude too high to carry any current of the proper magnitude. In fact, if screen 1 is removed completely, the reflection coefficient as obtained by the PMM program is less than 2 dB higher than with screen 1 in place.

Further, we observe that the cell sizes of the two CA screens are about $1.0\lambda_0$ and $1.4\lambda_0$. Such large interelement spacings result in numerous grating lobes, which at some frequencies and angle of incidence will coincide with the backscatter and specular directions. In general, they can add significantly to the bistatic reflection and should therefore be avoided at all cost. Although they are reduced as the result of the excessive high surface resistance of the CA sheets, as pointed out above, they are still at a dangerously high level. In fact, cell sizes should in general be less than about $0.5\lambda_0$ for all FSS and phased array designs, as demonstrated in this chapter.

In view of the fact that both FSS screens are essentially ineffective, one may ask the logical question: How do we explain the absorption? The answer is simply: It came from all the lossy dielectric sheets. They simply create a multilayered Jaumann absorber. In fact, both of the papers referred to above [5,6] contain extensive identical tables of ε_r' and ε_r'' for 10 different dielectric materials as a function of frequency. Their values are $4.48275 < \varepsilon_r' < 24.77497$ and $1.84577 < \varepsilon_r'' < 18.98354$ and are given in steps of 1 GHz from 19 to 36 GHz. Each of the two MGA designs contains eight dielectric slabs, some lossy and some essentially lossless. However, comparison with the designs in this chapter suggests that although it is possible to make materials with these dielectric constants, these precise values are not necessary, nor do we need that many layers in the present case.

Finally, it should be noted that many other papers [7-9] have considered absorbers similar to the types considered here. However, since they are basically narrowbanded and designed for normal angle of incidence only, it is not pertinent to compare them with the designs above.

4.7   CONCLUSIONS

In this chapter we first presented an absorber suggested by Lagarkov and Kisel. It can be described as a resistive screen backed by a flat lens backed by a ground plane. The purpose of the flat lens was to make the effect of the ground plane independent of the angle of incidence. It did indeed do that, but was hampered by the fact that a working flat lens just was not available. Another shortcoming was that the resistive sheet was not scan compensated. As a consequence, no actual calculated results were shown.

True to the concept for this book, we proceeded to present alternative, realistic designs. More specifically, we investigated absorbers made of resistive sheets placed in a stratified dielectric medium. We started with the simple Salisbury screen as a baseline, progressed to the Jaumann, and finally, to more sophisticated circuit analog absorbers made of resistive sheets provided with FSS features. Although such designs have been the subject of rather intense investigations for a normal angle of incidence [2, Sec. 9.5.4], what is new in this chapter is that we also find the specular scattered field for oblique incidence and arbitrary polarization. We found in all cases that a strong shift upward in frequency occurred with increasing angle of incidence. This was due primarily to the fact that the electrical spacing between the ground plane and the resistive sheet is given by $\beta_1 d_1 r_{1y}$, where r_{1y} is the cosine to the angle of incidence inside the dielectric, if any. As the angle of incidence increases, r_{1y} will become smaller, and consequently, β_1 must increase to maintain $\beta_1 d_1 r_{1y} \sim \pi/2$. The fundamental cure for this dilemma was simply to fill the space between ground plane and resistive sheet with a dielectric. As a result of Snell's law, r_{1y} will be smaller with higher ε_r and thereby provide greater stability with angle of incidence. However, we also observed some reduction in bandwidth. Thus, the dielectric constant ε_1 should not be increased indiscriminately. Values around 2 to 3 seem to be a good compromise. We also observed a mismatch between the absorber sheets and the incident field that increased with angle of incidence. This phenomenon is the same as that observed for matched arrays scattering in the bistatic direction when exposed to an incident plane wave, discussed extensively earlier. Thus, we only had to demonstrate the simplest cure, the use of a dielectric matching slab in front. For ideal scan compensation it was found that the dielectric constant of such a plate should be approximately $\varepsilon_r \sim 1 + \cos\theta_{1i}$, where θ_{1i} is the maximum angle of incidence in air. We found this to be a good choice.

However, what is new is that we also found that use of dielectric matching slabs could be used to increase the bandwidth significantly. The proper dielectric constant for scan compensation could also yield a large bandwidth by proper choice of the resistance of the resistive sheets, with or without the FSS feature. In fact, the relative bandwidth could typically increase from $f_H/f_L \sim 1.2$ for a simple Salisbury screen to about 1.4 for parallel and 2.0 for orthogonal polarization for a single resistive layer provided with a dielectric matching plate. For a single-layer CA sheet with dielectric matching plate we found the relative bandwidth to be equal to $f_H/f_L \sim 2$ for parallel and equal to $f_H/f_L \sim 2.7$ for orthogonal polarization (angle of incidence is $45°$). Note that none of these designs has been optimized. Inspection of the appropriate Smith charts (Figures 4.11 and 4.12) shows that further improvement is possible: for example, by a slight relocation of curve ④. A most interesting comparison was made with designs obtained entirely by numerical optimization: namely, the microgenetic algorithm [5,6].

REFERENCES

[1] W. W. Salisbury, Absorbent body of electromagnetic waves, U.S. patent 2,599,944, June 10, 1952.

[2] B. A. Munk, *Frequency Selective Surfaces: Theory and Design*, Wiley, New York, 2000.

[3] B. A. Munk, *Finite Antenna Arrays and FSS*, Wiley, Hoboken, NJ, 2003.

[4] B. Monacelli, J. Pryor, B. A. Munk, D. Kotter, and G. D. Boreman, Infrared frequency selective surface based on circuit-analog square loop design, *IEEE Trans. Antennas Propag.*, vol. 53, pp. 745–752, Feb. 2005.

[5] S. Chakravarty, R. Mittra, and N. R. Williams, On the application of the micro-genetic algorithm to the design of broad-band microwave absorbers comprising frequency-selective surfaces embedded in multilayered dielectric media, *IEEE Trans. Microwaves Tech.*, vol. 49, no. 6, pp. 1050–1059, June 2001.

[6] S. Charkravarty, R. Mittra, and N. R. Williams, Application of microgenetic algorithm (MGA) to the design of broad-band microwave absorbers using multiple frequency selective surface screens buried in dielectric, *IEEE Trans. Antennas Propag.*, vol. 50, no. 3, pp. 284–296, Mar. 2002.

[7] D. J. Kern and D. H. Werner, A genetic algorithm approach to the design of ultra-thin electromagnetic band-gap absorbers, *Microwave Opt. Technol. Lett.*, vol. 38, no. 1, July 2003, pp. 61–64.

[8] N. Engheta, Thin absorbing screens using metamaterials surfaces, *IEEE International Symposium on Antennas and Propagation*, San Antonio, TX, June 16–21, 2002.

[9] H. Mosallaei and K. Sarabandi, A one-layer ultra-thin meta-surface absorber, *IEEE International Symposium on Antennas and Propagation and USNC/URSI National Radio Science Meeting*, Washington, DC, July 3–8, 2005.

5 The Titan Antenna: An Alternative to Magnetic Ground Planes

5.1 INTRODUCTION

The Titan antenna will probably never be built again—so why talk about it?

> First: This was the most difficult matching problem I ever encountered in my entire career. Consequently, it contains many "neat tricks" that could be useful for antenna designers in general.

> Second: The Titan is one of only a handful of antennas operating at two frequency bands (140 to 150 and 250 to 270 MHz) using the same antenna elements and the same matching networks for both bands.

> Third: The distance between the mast and dipole elements was only about 0.015λ at the low band (the antenna was to fit inside a 12-inch silo when stored).

> Also: It would be an excellent illustration of practical antenna design for both students and professors. As pointed out by my good friend Dr. Ruth Rotman, it is an example of an antenna that could not have been developed just by computation. You must know how to use a Smith chart.

> Finally: It shows an alternative to the magnetic ground plane.

The purpose of the Titan antennas was to transmit data between missile sites. There were several identical vertical polarized antennas, each stored in its own silo at each site. When an antenna was activated, it would be raised out of the silo driven by compressed air. When an antenna was knocked out, another antenna would be activated automatically.

The mast consisted of two steel pipes with diameters of 7 and 6 inches welded together in the middle. For mechanical reasons these dimensions

Metamaterials: Critique and Alternatives, By Ben A. Munk
Copyright © 2009 John Wiley & Sons, Inc.

could not be reduced. Further, for economical reasons the diameter of the silo could not be more than 12 inches. Add to that a radome with a wall thickness of about 0.25 inch that should fit between the antenna and the silo. Cutting out the details, the distance from the center of the dipole element to the surface of the mast could be no greater than about 1.2 inch or about 0.015λ at the low frequency. That is the smallest spacing between a dipole and a ground plane that I have ever seen. It is often believed that this situation would require a magnetic ground plane. However, the solution given here is far superior.

5.2   LAYOUT OF THE ANTENNA

Back-of-the-envelope calculations quickly revealed that four dipoles placed on one side of the mast would be sufficient to produce the desired gain in the low band. This would also produce a desirable cardioid pattern in the horizontal plane. However, since there was no room for another set of dipoles at the high band, the radiation simply would have to take place from the same elements as well.

The most obvious solution to this problem is shown in Figure 5.1. In part (a) we show two of the four dipoles and their electrical dimensions at the low band. Similarly, in part (b) we show the same two dipoles with electrical dimensions at the high band. At the low-frequency band we observe an interelement spacing of $0.8\lambda_L$, which in that range will produce almost maximum gain. However, at the high band we observe that the interelement spacing is $1.45\lambda_H$. This spacing will lead to formation of grating lobes with subsequent loss of gain that could not be tolerated.

This dilemma was solved by using sleeve dipoles, as shown in Figure 5.2, instead of simple dipoles. Let us first consider the high band at the bottom of the figure since this is primarily what drives the design at this point. We observe that the sleeve dipole is comprised of a midsection of length $0.2\lambda_H$ and two outer sections each of length $0.4\lambda_H$. These dimensions were chosen such that the current distribution as shown in the figure was considerable larger at the two outer sections than for the midsection. In fact, the current distribution at the high band looked almost like two half-wave dipoles with end-to-end separation of $0.2\lambda_H$. Similarly, the tip-to-tip distance of the original dipoles had been reduced from $0.55\lambda_H$ in Figure 5.1 to $0.45\lambda_H$ in Figure 5.2. Certainly, these interelement spacings were not ideal, but as we shall see, were sufficient to reduce the grating lobes to acceptable levels. Keep in mind that the dipole dimensions were also affecting the impedance critically, as will be seen later.

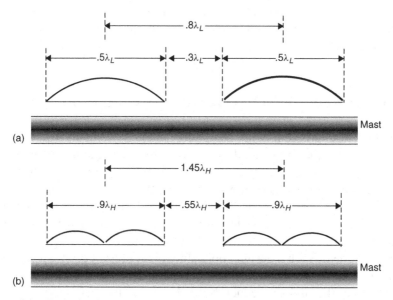

Figure 5.1 Typical current distribution on simple dipoles: (a) at the low band; (b) at the high band. Note that the interelement spacing at the high band is $1.45\lambda_H$, leading to grating lobes and unacceptable loss of gain.

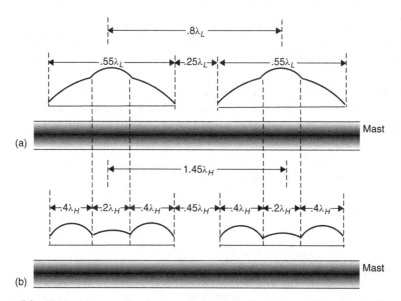

Figure 5.2 Typical current distribution on sleeve dipoles: (a) at the low band; (b) at the high band. Note that the interelement spacing at the high band is the same as before, but the current distribution now looks almost like two halfwave dipoles with tip-to-tip spacing of $0.2\lambda_H$ (instead of zero), and the gap between the sleeve dipoles has been reduced to $0.45\lambda_H$ instead of $0.55\lambda_H$. Still not ideal, but acceptable.

Finally, we note that the current distribution at the low band, as shown in Figure 5.2a, is perfectly ideal. No gain problems were ever experienced in that band.

5.3   ON DOUBLE-BAND MATCHING IN GENERAL

A fundamental problem with designing antenna elements for two frequency bands is that, in general, it is very difficult to change the elements such that only one frequency band is affected. For example, if the length of the element is changed, it will, in general, affect the impedance in both bands, and what is desirable for one band may be just the opposite for the other. Thus, one should try to design elements such that certain adjustments affect only one band and not the other.

An example of such a design applicable to out present problem is shown in Figure 5.3. We show here both the current and voltage distribution along an element at the high-frequency band. Note that the voltage distribution has a null approximately in the middle of the two outer sections. That means that we can attach whatever at that location without changing either the current or the voltage distribution or, more precisely, the impedance at the high-frequency band. We now attach two slanted conducting straps, each with one end connected to the mast and the other two ends to the voltage null point, as shown in Figure 5.3. The outer ends, connected to the neutral points, can be moved somewhat up and down. Although this will have only a minor effect on the impedance in the high-frequency band, the two slanted sections have turned the element into something like a folded dipole at the low band (actually a Γ-match), where the impedance changes significantly when the slanted straps are moved up and down.

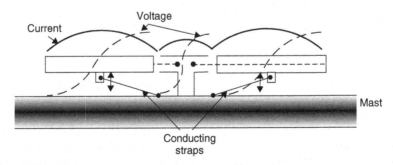

Figure 5.3   Current and voltage distribution at the high band on a sleeve dipole. Where the voltage is zero, we can attach conducting straps essentially without disturbing the fields in the high band.

Incidentally, the easiest way to locate the voltage null along the elements at the high band is probably experimentally. You simply record the terminal impedance at the center frequency of the high band. Next, you slide your fingers along either of the two outer dipole sections. When the impedance does not change, you have located the voltage null point quite precisely and you ask the machinist to attach the strap holders just there. Next time you update your résumé, you may even claim "hands-on experience"!

5.4 MATCHING THE SLEEVE ELEMENTS

It would be nothing short of naiveté to assume that the array of sleeve dipoles could be matched to anything close to 50 Ω by simple adjustment of the straps and the element dimensions in general. Ideally, the distance between the dipoles and the mast should be considerably larger than $0.015\lambda_L$, but that was not an option, as explained earlier. In order to understand the rather complex matching scheme for this antenna, let us consider the schematic of a single sleeve dipole, as shown in Figure 5.4. The impedances at the two feedpoints marked ① are obtained by measurement of the impedance at the terminals marked ④. They are shown in the Smith chart in Figure 5.5 and denoted L① and H① for the low-and high-frequency bands, respectively. For a refresher on matching, in general, see Appendix B in ref. 1.

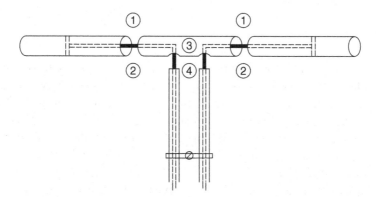

Figure 5.4 Schematic of a sleeve dipole and the "reversing" reactance ② at the high band: a coaxial stub inside the outer dipole sections. ④ is a balanced transmission line formed by the outer conductors of two 50-Ω coaxial cables joined by a conducting clamp some distance below the feeding point ④.

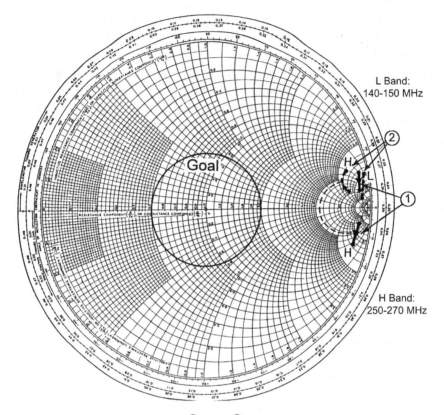

L Band:
140-150 MHz

H Band:
250-270 MHz

Figure 5.5   The impedance curves L① and H① (low and high bands, respectively) right at the feedpoints ① of the sleeve dipole (see Figure 5.4). Also shown are the curves L② and H②, which include the coaxial reactances at ②. Task: To transform L② and H② to the inside of the goal circle (VSWR = 2).

The goal is to transform both of these impedance curves to the inside of a circle with VSWR = 2 by using the same matching network. Considering the relatively small size of the two impedance curves L① and H①, this seems a trivial matter, at least to the untrained eye. However, two facts must be kept in mind. First, the nature of the Smith chart is such that curves at the edge will be greatly enlarged when transformed merely mathematically to the center region. However, what is more crucial is how the actual transformation is executed. For example, if we use a conventional single-stub tuner, we will quickly find that although the center frequency could be transformed to the center of the Smith chart, the lower and higher frequencies of the band would be completely outside our "goal circle."

An excellent example of this situation is shown in Figure B2 in ref. 1. Also discussed there is what to do about it: namely, "reverse" the impedance curve, which means that it should run "backward" in the Smith chart.* This is typically done by insertion of, for example, a pure reactance, either in series or in parallel with the pertinent antenna impedance. An example of this technique is shown in Figure B3 of ref. 1. Note that the final VSWR at 50 Ω is now less than 1.1. (No formal optimization was used to obtain this result, only a parametric study.)

In the present case we have solved the "reverse" problem by creating a series coaxial stub by inserting an inner conductor into part of the two outer sections of the sleeve dipoles as illustrated in Figure 5.4. The two coaxial stubs are short circuited at the outer ends, and their electrical length is slightly shorter than a quarter wavelength at the high band. Thus, their input impedances are purely inductive such that the high-frequency band H① is transformed up into curve H② as illustrated in Figure 5.5 (for details, see Appendix B in ref. 1). The remarkable feature of H② is that it has the higher frequencies located above the lower such that it runs backward when seen from the center of the Smith chart. That means that the curve H② is compressed in frequency when connected to a cable of proper length and characteristic impedance, as illustrated shortly.

We further note that the low band L① in Figure 5.5 is being moved into curve L② by the inductive stubs. However, this transformed curve is not reversed, due to the fact that L① is located in the inductive part of the Smith chart. In fact, L② is unfortunately being moved toward the rim of the Smith chart, but by no more than we can control, due to the fact that the inductive reactance of the two inductive stubs is fairly low at the low frequencies.

The actual reversing of the low-frequency band will not take place before the curves L② and H② are transformed into L③ and H③, respectively, at the terminals ③, as shown in Figure 5.4. The actual transformed curves are shown in the Smith chart in Figure 5.6. Note that both of these curves are now located in the capacitive part of the Smith chart. We need to reverse L③, whereas H③ essentially should be left alone. What kind of impedance would be capable of doing just that, and where would it come from physically?

We get the answer to that question if we again study the feeding arrangement in Figure 5.4. We observe that the sleeve dipoles are fed by two 50-Ω coaxial transmission lines operating as a balanced 100-Ω

---

*Running backward in the Smith chart does not violate Foster's reactance theorem, which pertains only to pure reactances located entirely on the rim of the Smith chart, not to impedances inside the chart. (Sure, we are very close to the edge, but as we will see, far enough inside to be "saved.")

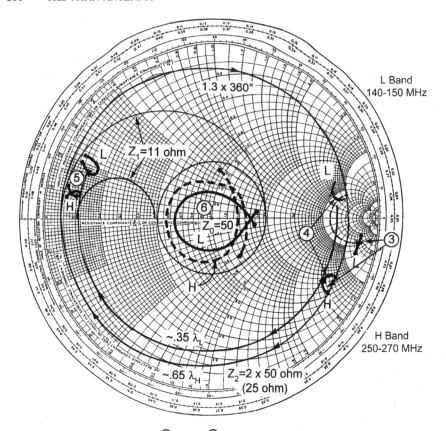

Figure 5.6 The two curves L② and H② (see Figure 5.5) are transformed to the terminal area ③ and denoted L③ and H③, respectively. We then add the reactance from the balanced stub and obtain L④ ahd H④. Further transformation takes place along the balanced transmission line, yielding L⑤ and H⑤ (see Figures 5.7 and 5.8). Next, transformation takes place in two coaxial sections, each with characteristic impedance 22 Ω yielding the curves L⑥ and H⑥ at a 200-Ω level. Connecting adjacent dipoles in parallel yields a 100-Ω level that matches right into the balanced 100-Ω transmission line. The two antenna halves are then connected in parallel at the balun to match right into the 50-Ω main line.

twin-lead line. Normally, the two outer coaxial shields should be connected to each other at ④ by a conducting clamp or simply soldered together. However, we have left it open and instead connected the two outer shields by a clamp at a lower position, as shown in Figure 5.4. These two shields act on the outside as a section of a twin-lead cable short circuited at the lower end by the clamp. Thus, the input impedance of the twin-lead stub at ④ is purely reactive and connected in series with the impedances L③ and H③. If the distance from the lower clamp to ④ is somewhat shorter than a quarter-wavelength at the lower-frequency band, the twin-lead impedance at ④ will be purely inductive, such that the impedance curve L③ will be transformed into the inverted curve L④ shown in Figure 5.6.

However, for the outer twin-lead section not to affect the impedance H③ significantly (since it is already reversed), we want its input impedance at ④ to act electrically as a short at the high band (i.e., the distance from the lower clamp to ④ should ideally be one half-wavelength at the high frequency). Typically, the electrical length at the center of the low band would be about $0.2\lambda_L$; that is, the electrical length at the center of the high band would be approximately $(0.2)(260/145) = 0.3\lambda_H$ (i.e., significantly shorter than $0.5\lambda_H$).

This dilemma was solved by placing a shunt capacitor across the outer twin-lead line approximately midway between the lower clamp and ④. This capacitor is not shown in Figure 5.4 but can be seen clearly in the physical layout shown in Figure 5.7. Such a capacitor would make the electrical length of the twin-lead stub larger at both the low and high frequencies. However, it will affect the high frequency most because the capacitor is located close to a voltage maximum at the high frequency and not as close at the low frequencies. The detailed calculation (in a Smith chart, of course!) is left as an exercise for the student. Note further that the electrical short at ④ does not have to be terribly effective since the impedance level at H④ is so high. Thus, H③ is essentially identical to H④.

Finally, we show in Figure 5.7 a rather detailed picture of what an individual sleeve dipole actually looked like. We note the coaxial matching stubs inside the outer dipole sections as well as the two 50-Ω coaxial cables working as a balanced feedline with the shorting bar and the capacitor, discussed earlier. Finally, note the three robust cable holders welded to the mast. Why three and not just two? Well, that's an exciting story that we tell in Section 5.8.

5.5 FURTHER MATCHING: THE MAIN DISTRIBUTION NETWORK

We saw above how the impedance curves L④ and H④ shown in Figure 5.6 got reversed. This step should always be the first and be performed as close to the element terminals as possible before the impedance curves start to "curl up" with increasing frequency. However, further matching takes place in the main feedwork shown in Figure 5.8. We start with the

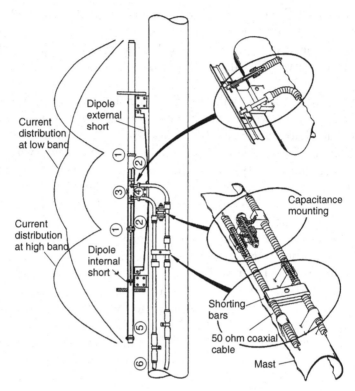

Figure 5.7   Detailed drawing of a single-sleeve dipole being fed from a balanced line comprised of two 50-Ω coaxial cables. Also shown are the two coaxial sections with characteristic impedance 22 Ω, each located between ⑤ and ⑥. Finally, note the three cable holders welded to the mast, as well as the capacitance for adjusting the length of the balanced stub.

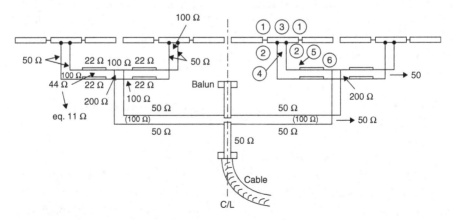

Figure 5.8   Complete schematic of the feed network for the entire antenna array.

two 50-Ω balanced coaxial lines connected directly to the terminals ④ of each sleeve dipole. The other ends are connected to a pair of specially machined coaxial sections each with characteristic impedance 22 Ω. Their endpoints are denoted ⑤ and ⑥, as indicated in Figure 5.7 and to the right in Figure 5.8. In Figure 5.6 we see how the impedance curves L④ and H④ are being transformed into L⑤ and H⑤, respectively, by the balanced coaxial transmission line. We observe that these impedance curves are located in the same neighborhood in the Smith chart despite the fact that the electrical length of the balanced coaxial line is $0.35\lambda_L$ at the low frequency and $0.65\lambda_H$ at the high frequency, respectively.

Next, L⑤ and H⑤ are transformed into L⑥ and H⑥, respectively, via the two 22-Ω sections. We note that this transformation was originally done the usual way in a Smith chart normalized to 22 Ω. However, we like to show the transformation as shown in Figure 5.6 because it shows so vividly how an impedance curve is greatly enlarged when it is transformed from the outer to the inner part of the Smith chart, as discussed earlier.

We note further that L⑥, as well as H⑥, are at the 200-Ω level such that when two adjacent dipole elements are connected in parallel, the level is reduced to the 100-Ω level. Thus, it is matched directly into the last section of a balanced 2 × 50-Ω line. Note further that since no frequency-sensitive transformers are involved, there are no bandwidth limitations except for the bandwidth of the curves L⑥ and H⑥ themselves.

Finally, the two 100-Ω balanced transmission lines for each antenna half are joined in parallel at the balun as shown in Figure 5.8. That brings the final level down to 50 Ω, which is the level used in Figure 5.6 when showing the two curves L⑥ and H⑥, transformed to the balun.

5.6 THE BALUN

It was pointed out earlier how the feed network was designed without any frequency-sensitive parts. Unfortunately, it did not seem possible to pull similar tricks in the case of the balun. We simply needed a balun with a bandwidth of about 3 : 1. Finally, it was decided to use the design shown in Figure 5.9. This layout, often appropriately called a *mast balun*, may be compared with the well-known *Roberts balun*, the main difference being that whereas the Roberts balun is oblong, the mast balun is flattened out. (For an in-depth discussion of baluns in general, see Chapter 23 in ref. 2.) Both of these designs will balance almost perfectly at all frequencies. Their frequency limitation is due to the reactance of the stubs, consisting of the outer coaxial shields and the surface of the mast. Since the

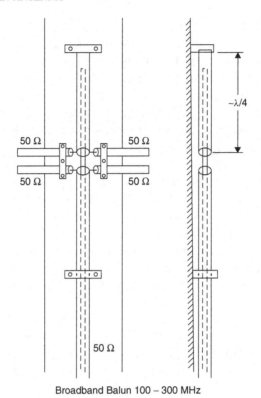

Broadband Balun 100 – 300 MHz

Figure 5.9 Broadband balun mounted on the outside of the mast. The open-ended stub inside the top section becomes capacitive at the lower frequencies, whereas the two balun shunts become inductive. That increases the bandwidth substantially.

characteristic impedances of these two stubs are fairly large and the two reactances on each side are in series, we will end up with a fairly large total reactance, in parallel with a relatively low (ca. 50-Ω) impedance. Further, a compensating stub was inserted in the top balun. Thus, a bandwidth with a 3 : 1 frequency range was readily obtained. We finally observe in Figure 5.9 how the two balanced 100-Ω transmission lines to each array half are connected in parallel at the 50-Ω balun, creating perfect broadband matching.

5.7 THE RADIATION PATTERN

Typical radiation patterns in the horizontal plane are shown in Figure 5.10 at the low and high bands, respectively. The cardioid patterns are due

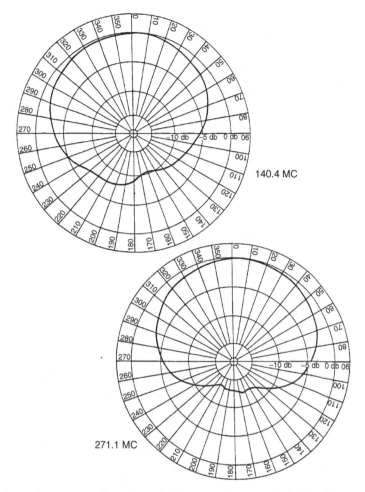

Figure 5.10 Measured horizontal cardoid pattern at the low and high bands, respectively.

entirely to the shadowing effect of the mast and had the desired shape. A few specimens were made with a corner reflector added to enhance the gain (see Figure 5.15). The effect on the impedance from the reflector was minimal (more than can be said about the effect from the mast).

Similarly, we show in Figure 5.11 the vertical pattern in the low and high bands. The low pattern is quite satisfactory, while the high band has some small grating lobes, as anticipated (see Section 5.2). However, since there was no sidelobe specifications and we met the gain requirement, it was of no concern.

The antenna was measured in the horizontal position on a primitive turntable that could handle the approximate 1-ton weight, as required. It

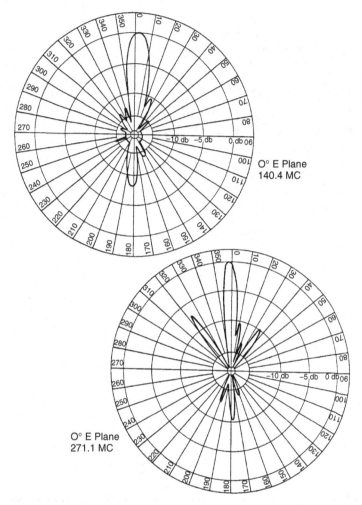

Figure 5.11 Measured vertical pattern at the low and high bands, respectively. The grating lobes, visible at the high band, are sufficiently suppressed not to affect adversely the gain in the high band.

was being pushed around by my technician, Jim, using the "Armstrong" method. That explains the slight misalignment of the pattern.

5.8 SOMETHING THAT SOUNDS TOO GOOD TO BE TRUE USUALLY IS

When the first prototype of the Titan antenna was tested, I expected the VSWR to slightly exceed 2 : 1. It was hoped that minor adjustments could

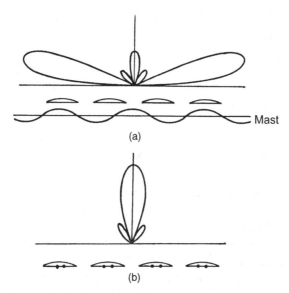

Figure 5.12 (a) Undesirable pattern caused by mast oscillation (the entire mast was "hot"). (b) Vertical pattern desired For an explanation, see the text.

bring it within specs. Much to my surprise, the VSWR was considerably lower, actually 1.6. I was immediately suspicious—and sure enough, the vertical antenna pattern looked like a disaster. Instead of a pattern as indicated in Figure 5.12b, it looked like the one shown in part (a). Very quickly it was suspected that the entire mast was being excited, as indicated in the figure. In fact, when the impedance was measured, it would change dramatically if you placed your hand on the end of the mast. In other words: It was "hot"!

Obviously, some imperfect balance got the entire mast excited more than the dipoles. But where did the imbalance come from? The first suspect was, of course, the balun, However, it seemed okay. After careful reflections over what was different between the lab model and the first prototype, it was established that the 100-Ω balanced line attached directly to the dipole terminals ④ had hung down from the dipoles at 90° from the mast for the lab model (the mast was placed in a horizontal position). This was convenient when working on matching in general. In contrast, the mechanical engineers had bent the feed cables so that they were parallel with the mast as shown in Figure 5.7 for the prototype, and welded the two lower cable holders solidly to the mast (all with my approval, of course). However, note that one of the two 50-Ω cables was longer than the other, due to a larger bending radius (only on the outside; inside, the cable lengths were cut to the correct length). There was the imbalance.

More precisely, when a short was placed from the longest outside cable to the mast at the right point, the patterns were perfect. The mechanical engineer was not very happy about moving one cable holder to the new position closer to the dipoles. I said: "Just leave it and put another where it gives the right balance." That solution was fine with him—and that is why we have three cable holders and not just two!

5.9   EFFICIENCY MEASUREMENTS

Every time a conducting element is placed close to a ground plane or in this case a substantial mast, the terminal impedance will be driven toward the rim of the Smith chart. Consequently, the antenna becomes more susceptible to ohmic losses in the elements. Due to the complexity of the present antenna, the sponsor determined to get an idea about the efficiency simply by measuring the temperature rise at seven agreed-upon locations on an element when exposed to full power (1 kW). This was done by use of thermocouplers. Since all parts of the elements were grounded to the mast, it was simple to place the control wires such that they did not interfere with the radiated field. Although I do not recall the temperature specs, I vividly remember that the rise in temperature was substantially less than the specs. This evidently suggests that the ohmic losses are quite a bit smaller than is often assumed (see, e.g., Figs. 2–11 and 2–12 in ref. 2).

5.10   A COMMON MISCONCEPTION

A *magnetic ground plane* is, in general, defined as a structure with infinite high surface impedance, in contrast with an electric-conducting ground plane, where the surface impedance is zero. Thus, when using an electric ground plane, an electric current element must be placed some distance from the ground plane. In contrast, an electric current element can conceptually be placed directly on top of a magnetic ground plane.

Many elaborate schemes have been proposed in the literature for producing magnetic ground planes. However, they all follow the same basic design concept: An inductive impedance is created by placing an electric ground plane about 0.05 to 0.15λ behind the plane of the magnetic ground plane. You then add a shunt capacitance by adding either a dielectric slab or a periodic structure of conducting elements. This rather simple idea is being elevated to immediate respectability by denoting the latter concoction as belonging to the metamaterial family. Sure, it works—but only over a bandwidth of a few percent, depending on the distance to the electric ground plane. I have stated repeatedly that you are, in general, better off by using "smart" matching by placing your electric current element some distance from the electric ground plane (see, e.g., Sections 6.12.4 and 6.12.5 in ref. 1). The Titan antenna presented in this chapter is further proof of this statement. It not only provides an excellent match over about an 8% bandwidth in one band, but actually in two widely separate bands for a ground plane distance as short as 0.015λ. At that distance a magnetic ground plane would be able to obtain a bandwidth of only 1 to 2%, and in one band only! The trick is, of course, that we apply a matching technique that inverts the antenna impedance (smart matching). Is it any wonder that I am somewhat skeptical of "blind" optimization?

Why discuss this subject in this book? Because magnetic ground planes are perceived to be metamaterials.

5.11   WE PUT THE MAGNETIC GROUND PLANE TO REST

The concept referred to as the magnetic ground plane (MG) has been investigated for some time in several papers and books. In addition, several papers dealing with the MG have been rejected before publication simply because they missed the point or were just trivial (as were many of the published "peer-reviewed" articles).

The fundamental idea behind an MG is straightforward: It should provide us with a structure yielding infinite or at least a large sheet impedance, such that elements with electric currents will not be shorted out even when placed directly on top of the MG. Obviously, the image of the electric currents in the MG will always be in phase with the original. It is widely believed, or at least implied, that maximum directivity of a single element or column is obtained when the element(s) are placed directly on top of the MG. As illustrated in Figure 5.13, that is indeed the case. In Figure 5.13a we show the radiation pattern when the elements are placed directly on an MG, and in Figure 5.13b, when they are placed $\lambda/4$ in front of an MG. Obviously, the image and the original will add in phase in all directions and produce an omnidirectional pattern in the first case, while in the latter case, they will cancel each other in the forward direction (i.e., the directivity is zero in the forward direction). It is interesting to note that if we were to replace the MG with an electric ground plane, we would still obtain maximum directivity for "small" spacing but only a slightly smaller directivity for $\lambda/4$ spacing. In addition, the directivity for the electric ground plane is always higher than for the MG case, as discussed in Section 6.12.4 of ref. 1. So apart from the fact that magnetic conductors do not occur in nature, one wonders what is its claim to fame.

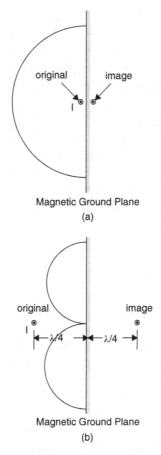

Figure 5.13 The radiation pattern of an electric current and a magnetic ground plane for (a) close spacing between the original and the magnetic ground plane; omnidirectional pattern. (b) $\lambda/4$ spacing; no field in the forward direction.

Let us consider how we may construct an artificial MG as illustrated in Figure 5.14. Typically, it consists of an electric ground plane in front of which we have placed our electrical elements. The spacing between these and the electrical ground plane may typically be 0.05 to 0.15λ (i.e., the impedance contribution from the electrical ground plane at the plane of the elements is inductive). Thus, if we place some form of capacitance next to the dipoles, the ground-plane impedance will be very large, as required of an MG. Such a capacitance can take many forms, ranging from a simple dielectric sheet to small conducting elements of almost any shape. Alternatively, we could load the electrical elements themselves with capacitors. Whatever we do, it is important to realize that the radiation

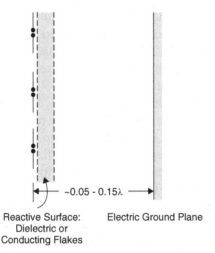

~0.05 - 0.15λ

Reactive Surface: Electric Ground Plane
Dielectric or
Conducting Flakes

Figure 5.14 An artificial magnetic ground plane (or high impedance) is typically created by placing a capacitive reactive surface a distance of about 0.05 to 0.15λ in front of an electric ground plane. The reactive surface can be a dielectric slab or a periodic structure of small conductors.

pattern of a large array is determined primarily by the highly pointed array factor, not by the element pattern. Thus, the directivity of the array is basically not affected by the spacing between the ground plane and the elements, nor by the capacitive loading. Thus, our only concern is how to match the total impedance observed at the terminals of the electrical elements. The value of the ground-plane impedance is irrelevant as long as the dipole impedance is not purely imaginary and different from zero and as long as matching is possible.

An example of an antenna following this design philosophy is given in Chapter 6 of ref. 1. Here we consider an array with a bandwidth of approximately 9 : 1 with a maximum VSWR of 2 : 1, considerably better than can be obtained with an MG. Another example is the Titan antenna discussed in this chapter, where the ground-plane spacing is merely about 0.015λ. The bandwidth of an MG for such a small spacing would be of the order of 1 to 2%. Note: A larger ground-plane spacing produces a larger bandwidth.

To summarize: The magnetic ground plane concept aims to produce a high impedance at the plane of the electrical elements by use of a dielectric slab or an equivalent. Alternatively and potentially much more effective is to combine the impedance of the electrical elements themselves with the ground-plane impedance to yield the proper total terminal impedance. If done cleverly, it may lead to a considerably larger bandwidth.

It is absolutely amazing how the simple magnetic ground plane concept has attained such importance and has resulted in numerous papers and book chapters. Most deplorable is the fact that most authors simply do not even consider the only really important factor: the terminal impedance of the antenna placed on top.

Every time a dielectric slab is placed next to a periodic structure, attention should be given to the possibility of surface waves. More on this subject and how to avoid it is may be found in my first book [3].

Typical periodic structures of conducting "flakes" can also be used to produce the necessary capacitance. The shape may be square, rectangular, circular, or hexagonal, just to mention a few. Sometimes they are made in the form of loops, and the inner part may be filled with more or less meaningful wire structures. The oddest I have seen was filled with some fractal contraption. Considering that the diameter of loops was only a small fraction of a wavelength, such a design seems rather pointless. Presumably, "smiley faces" would work just as well!

For a systematic presentation of a "high-impedance" ground plane, the reader is referred to Chapter 11 of ref. 4.

5.12   CONCLUSIONS

We designed an antenna with several unique features, such as:

1. The antenna operated in two bands, 140 to 150 and 250 to 270 MHz, using the same special element (a sleeve dipole) in both bands.
2. The same matching network was used for both bands.
3. The distance from the sleeve dipole to the surface of the mast was only $0.015\lambda_L$ at the low band, leading to extremely high terminal impedances in both bands.
4. The antenna fits inside a 12-inch silo and the diameter of the mast is 7 inches (for the bottom section). This explains the lack of room for more than one set of elements, as well as matching networks.
5. The main feeding system uses no frequency-sensitive components, yielding a bandwidth determined only by the elements per se.
6. A new type of broadband balun $(1:3)$, especially suitable to be mounted outside a mast, was developed.
7. Using magnetic ground planes is not only unnecessary but counterproductive.

    It has generally been agreed by everyone who analyzed this antenna that it constitutes one of the most sophisticated designs in

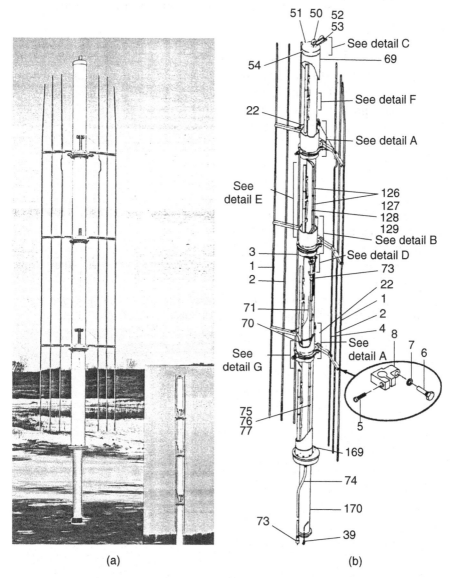

Figure 5.15 One version of the Titan antenna where a corner reflector has been added to the standard version. It folds up along the antenna when stored in a silo. It had only a minor effect on the impedance of the already strongly decoupled antenna impedances, due to the mast.

the VHF–UHF region. It is therefore suggested that the antenna be used as an illustration of antenna technology in general, and practical antenna design in particular, for the benefit of both students and instructors.

P.S. The complexity of this antenna often makes people believe that it would be too tricky to build. Actually, counting all the variants of this antenna, almost 200 antennas were build and installed without any significant problems. One version had a corner reflector that could be folded up along the mast to fit in the silo when stored as shown in Figure 5.15.

REFERENCES

[1] B. A. Munk, *Finite Antenna Arrays and FSS*, Wiley, Hoboken, NJ, 2003.

[2] J. D. Kraus and R. J. Marhafka, *Antennas for All Applications*, 3rd ed., McGraw-Hill, New York, 2002.

[3] B. A. Munk, *Frequency Selective Surfaces: Theory and Design*, Wiley, New York, 2000.

[4] N. Engheta and R. Ziolkowski, *Metamaterials: Physics and Engineering Explorations*, IEEE Press/Wiley-Interscience, Hoboken, NJ, 2006.

6 Summary and Concluding Remarks

6.1 BACKGROUND

In 1968, Veselago asked a perfectly good question: What would happen if the permeability μ and the permittivity ε for a material were both negative? [1]. He concluded that the refractive index between two media with μ_1, $\varepsilon_1 > 0$ and μ_2, $\varepsilon_2 < 0$, respectively, would be negative. He arrived at this result simply by matching the boundary conditions. A logical extension of this discovery led to the postulate of the flat lens (i.e., a perfectly flat slab with $\mu = \varepsilon = -1$). It was claimed that it could focus energy inside as well as outside the slab.

For almost 30 years after Veselago published his paper, it appears that it attracted almost no interest at all: that is, until Pendry came along in the mid-1990s and suggested that materials with μ, $\varepsilon < 0$ could be made artificially by use of periodic structures of various wire elements [2–5]. Not long after, at the turn of the century, a group of physicists at the University of San Diego built a prismlike configuration consisting of straight wires and split-ring resonators as suggested by Pendry [6–9]. They claimed that they had indeed observed negative refraction experimentally. It was followed quickly by an avalanche of papers, books, and conventions in which negative refraction was presented as a fact of life. Almost no papers with a critical view were being published.

Meanwhile, numerous configurations claiming negative μ and ε by the theorists were tested experimentally—by professionals, not just students who wanted to graduate! No negative μ and ε were ever observed, and negative refraction was typically so weak that it could be characterized as being simply spurious signals.

This author never had a contract concerning metamaterials, nor did I ever ask for one. However, I did have some experience in periodic structures in general. It therefore became quite common that in casual

Metamaterials: Critique and Alternatives, By Ben A. Munk
Copyright © 2009 John Wiley & Sons, Inc.

conversation I would hear remarks such as "Hey, Ben, this looks like your stuff. What do you think about it?"

I gave my first oral opinion at the Torino conference ICEAA01 in 2001 after a paper given by Engheta [10,11]. He was considering the resonance frequency of a cavity between two ground planes half-filled with material with μ, $\varepsilon > 0$ and half with μ, $\varepsilon < 0$, respectively. The resonance frequency remained almost constant from dc to broad daylight! I argued from the floor that if the entire cavity were filled with μ, $\varepsilon < 0$, the resonance frequency would move counterclockwise in a Smith chart, violating Foster's reactance theorem (see Figure 1.2). I added, to the great indignation of the audience, that although I found the paper interesting, I did not believe a word of it! However, so as not to jeopardize funding for any of my colleagues, I decided to put a hold on publishing a paper. Not before 2003 did I submit to *IEEE Transactions on Antennas and Propagation* a paper with the title "On Negative μ_1 and ε_1: Fact and Fiction." One reviewer found it interesting, whereas the two others were adamantly against it. However, rather than spending time fighting for publication, I decided to publish it some other time. It is given in Appendix A, five years after it was written, without editing.

Meanwhile, many potential sponsors were under intense pressure to support this new endeavor, which promised unbelievable returns. There were a few skeptics, but typically, they felt intimidated when confronted by the enormous output of publications that claimed to have observed a negative index of refraction. (How could so many be wrong? Besides, nobody wanted to miss out on the *big breakthrough*.) Essentially only one paper, by Valanju, Walser, and Valanju, objected strongly [12].

Many in government as well as industry wanted a publication that was rigorous but also emphasized what really went on from a physical point of view. Eventually, I decided to take a stab at it. The result is the book you hold in your hand.

6.2 THE FEATURES OF VESELAGO'S MATERIAL

Veselago concluded that essentially two features were associated with his material:

1. There was a negative index of refraction between two media with μ_1, $\varepsilon_1 > 0$ and μ_2, $\varepsilon_2 < 0$, respectively.
2. A plane wave propagating through his medium would be left-handed, meaning that $\bar{E}$, $\bar{H}$, and the direction of phase propagation form a left-handed triplet (see Figure 1.4) (the Poynting vector

still points in the same direction as for a normal right-handed wave propagating through a medium with μ_2, $\varepsilon_2 > 0$). It was later suggested that Veselago had notions about backward-traveling waves [13,14]. But there is no evidence to support that claim as far as this writer can determine.

It was later postulated by Pendry and others that:

3. A wave propagating through Veselago's medium would exhibit phase advance, in contrast to phase delay, when propagating in a medium with μ_2, $\varepsilon_2 > 0$ (see Figure 1.2).

4. The evanescent waves propagating through Veselago's medium would increase as they moved away from their origin.

It should be noted that claims 3 and 4 obviously are closely related from a theoretical point of view. Strangely enough, some people can accept one but not the other.

6.3   WHAT CAN A PERIODIC STRUCTURE ACTUALLY SIMULATE?

To the best of this author's knowledge, all attempts to realize Veselago's medium have so far been done by application of periodic structures using various elements. Considering the fact that we have some experience with periodic surfaces, it was therefore natural that we first investigated whether it was possible to simulate any feature of Veselago's medium by use of periodic surfaces using whatever elements we desired.

We concluded, with great certainty, that a negative index of refraction just was not possible when the structure was infinite and the interelement spacing less than $\lambda/2$. If the structure was finite and the interelement spacings still less than $\lambda/2$, weak radiation could be observed in the negative sector. It was, however, not negative refraction but merely radiation from either a surface wave characteristic of finite periodic structures or possibly simply a sidelobe from the main beam in the positive sector [15]. The weakness of these waves has nothing to do with either ohmic or dielectric losses. They are due merely to low radiation efficiency of surface waves in this particular case.

Further, if the interelement spacings were greater than $\lambda/2$, grating lobes could be observed. These are often confused with backward-traveling waves. However, there is absolutely nothing backward or left-handed about them. They are perfectly "normal" except that they can propagate antiparallel to the incident field.

We emphasize that these findings have absolutely nothing to do with the type of element that we use in the periodic structure. We concluded further, with great certainty, that all fields radiated from infinite periodic structures are right-handed regardless of element type and interelement spacings (see Section 1.6). Finally, we saw absolutely no evidence of evanescent waves with increasing amplitude or propagating waves with phase advance: that is, unless you terminate your structure in a resistive load. That has nothing to do with creating a new magic material but simply represents network tricks related to broadband matching. The author is quite familiar with this subject, as discussed in several sections of ref. [15] as well as in Chapter 5.

6.4 DID VESELAGO CHOOSE THE WRONG BRANCH CUT?

We stated above that a periodic structure could not realize a single feature characteristic of Veselago's material. It is therefore natural to revisit Veselago's original paper [1]. He arrived at his results by matching boundary conditions as usual. As mentioned in Chapter 1, he apparently did so correctly and formally postulated negative refraction. However, as also pointed out in Chapter 1, the solution was purely mathematical, and all such solutions must be checked for physical validity. We did just that by examination of the "perfect" lens (i.e., a flat slab with $\mu = \varepsilon = -1$). We found that in order to focus at all frequencies, the medium should provide us with a time advance, or "negative time," not delay as is normally the case for $\mu, \varepsilon > 0$.

From a mathematical point of view, negative time is just another parameter that is completely acceptable. However, from a physical point of view, it is not permitted. (If anyone out there has some negative time, please send me a slice; it could prolong my life!) See also Sections 6.5 and 1.11.

6.5 COULD WE EVER HAVE A NEGATIVE INDEX OF REFRACTION?

When Veselago postulated a negative index of refraction, he did not specify that periodic structures should be used. He left the creation of his material completely open; in fact, he was wondering whether there were good reasons why such a medium may not be realizable. It is therefore natural to ask the question: Could we ever create *any* material where a negative index of refraction could be possible? To answer this question,

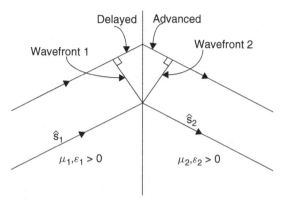

Figure 6.1 A wave propagating in direction $\hat{s}_1$ is incident upon a boundary between two media with μ_1, $\varepsilon_1 > 0$ and μ_2, $\varepsilon_2 < 0$. The upper part of the incident wave is delayed. Thus, the upper part of the refracted wave must be advanced when propagating in the direction $\hat{s}_2$ as shown (i.e., associated with negative time). This supports the findings by Valanju et al. [12].

consider Figure 6.1. It shows the boundary between two semi-infinite media with μ_1, $\varepsilon_1 > 0$ and μ_2, $\varepsilon_2 < 0$, respectively. A plane wave propagating in the direction $\hat{s}_1$ is incident upon this boundary. We ask: Is it possible to obtain a plane wave in the medium μ_2, ε_2 propagating in the direction $\hat{s}_2$, as shown in the figure? We observe that the upper part of the incident wavefront in medium 1 is delayed. Thus, the upper part of the wavefront in medium 2 must be advanced in time to create a plane wave as shown. In other words, medium 2 must somehow be associated with negative time. And as discussed earlier, that is not an option.

However, not everyone is convinced by the argument given in conjunction with Figure 6.1. Thus, in Figure 6.2 we give a somewhat more detailed explanation of why a flat lens just will not work even if negative refraction were possible. We consider a source in medium 1 emanating a single pulse along each of three (or more) rays denoted 0, 1, and 2, respectively. If we assume for the moment that negative refraction, $n_{12} = -1$, is possible, these rays will meet at the point denoted crossover 1 in the figure. Since ray 1 has a total path length $2\Delta l_1$ longer than that of ray 0, the pulse associated with ray 1 will arrive at the crosspoint with a time delay proportional to $2\Delta l_1$. Similarly, the pulse associated with ray 2 will be delayed by a time proportional to $2\Delta l_2$. Thus, all the pulses going through the crossover point will do so at different times and never be able

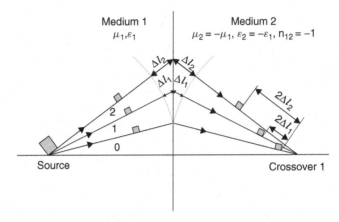

Figure 6.2 How a pulse starting at the source will arrive at crossover 1 with time delays depending on the inclination of the individual rays. Thus, a strong picture of the source at crossover 1 is not possible: or a flat lens does not work.

to reproduce a strong and undistorted reproduction of the "picture" at the source.*

This discussion showing the failure of the flat lens could actually stop right here. However, one could argue that it could "work" if time delay associated with positive time could be replaced with time advance associated with negative time, as suggested in the discussion of Figure 6.1. Thus, we must again conclude that Veselago's solution, although mathematically correct, is deficient in the physical world because it relies on negative time. Or, alternatively: the rotation in the counterclockwise direction in Figure 1.2 is not possible.

6.6 COULD VESELAGO HAVE AVOIDED THE WRONG SOLUTION?

It is interesting to note just exactly what boundary condition is responsible for producing a negative index of refraction. In fact, it is the normal component of the displacement field that is continuous when going from medium 1 to medium 2:

$$\bar{D}_1 = \varepsilon_1 \bar{E}_1 = \varepsilon_2 \bar{E}_2 = \bar{D}_2 \tag{6.1}$$

*This train of pulses can, of course, be decomposed into a spectrum of sinusoidal waves. Some of these may very well (actually quite likely) be in phase at the focal point. However, most of the components are not. Thus, our conclusion above that the crossover point is *not* a focal point was correct.

Obviously, when ε_1 and ε_2 have different signs, so will $\bar{E}_1$ and $\bar{E}_2$. And, as illustrated in Figure 1.20, that is precisely what formally leads to a negative index of refraction. However, as pointed out in Appendix A, Maxwell's equations can be written in the alternative form

$$\nabla \times \bar{E} = -j\beta_1 Z_1 \bar{H} \qquad (6.2)$$

$$Z_1 \nabla \times \bar{H} = j\beta_1 \bar{E} \qquad (6.3)$$

where, as usual,

$$\beta_1 = \omega\sqrt{\mu_1\varepsilon_1} \qquad (6.4)$$

$$Z_1 = \sqrt{\frac{\mu_1}{\varepsilon_1}} \qquad (6.5)$$

From equations (6.4) and (6.5) we see that neither β_1 nor Z_1 will change if μ_1 and ε_1 are both negative [i.e., (6.2) and (6.3) are unchanged as well and "nothing" happened]. Of course, we could have chosen the other branch cut in (6.4). However, this is trivial since it merely constitutes an identical wave propagating in the opposite direction. As to choosing Z_1 negative, everyone agrees that it is only relevant close to a "black hole" in outer space, not in a passive medium. See also equation (6.6).

Interestingly enough, Veselago specifically stated that to "learn something new about materials in general" we should use Maxwell's equations containing μ_1 and ε_1, as is generally done [1]. We did indeed learn many new startling features, most notoriously, negative index of refraction. However, we now know that this was a nonphysical solution that violated the physical laws, as illustrated, for example, in Figures 6.1 and 6.2 (see also Sections 1.11, 6.5, and 1.12.4).

6.7 SO WHAT CAME OUT OF IT?

As stated earlier, we did not see any convincing experiment showing a negative index of refraction (too weak a signal). Nor did we ever produce a flat lens, although numerous front-page pictures have been shown. We did get an avalanche of enthusiastic papers and at least four books, not counting this one, that can hardly be characterized as anything but critical [17-20].

Rarely did we see any papers in opposition. We can only speculate on the reasons for that. Undoubtedly, there were more believers than nonbelievers; and many of the skeptics were simply too timid to speak up. But most of all, getting an opposing paper published was almost impossible. It would eventually be rejected by the vast number of earlier authors and reviewers.

6.8   IS PUBLISHING THE ULTIMATE GOAL IN SCIENTIFIC RESEARCH?

Someone was once giving a very upbeat assessment of the future of metamaterials with a negative index of refraction. He finished jubilantly with

> *We saw 250 papers in 2004.*
> *In 2005 we expect even more!*

Well, sir, that is beautiful poetry indeed! But is the quantity of papers really a measure of the quality and significance of the research in question? I don't think so. Certainly, in the academic world "publish or perish" requires a minimum number of publications. That was not always the case. In fact, some of the greatest innovators wrote relatively few papers.

6.9   WHAT EXCITES A SCIENTIST?

It should be obvious by now that this author has the greatest respect for the science of metamaterials. Only when fundamental laws of physics are violated, in particular when a negative index of refraction is claimed, must I strongly object. Further, I see nothing wrong in Veselago asking his now famous question. Obviously, I think he made an honest mistake with regard to negative refractions and left-handedness based merely on boundary conditions. However, it would all have been inconsequential had it not been for Pendry, who apparently believed that merely producing a material with negative μ and ε would indeed result in negative refraction. He added more features, in particular that evanescent waves would increase as you moved away from the source leading to a superlens.

Although these concepts raised eyebrows in some camps, it would probably be correct to state that the vast majority of scientists really believed these new exciting ideas. This statement is based on the fact that we saw more than 1000 enthusiastic papers and hardly any critical ones. Further, technical conferences of highest standards were customarily dominated completely by metamaterials. Mostly, academics were involved, although some activity was detected in industry. Typically, however, they were observing others and not spending their own IRAD money. However, there was plenty of funding from the government. And undoubtedly, that was one of the primary drivers. In fact, being a scientist today makes you face many difficult choices.

But what really fascinates me is how such a misconception could live for more than a decade. The fact is that although everybody knew Veselago's name and that he postulated a negative index of refraction, very few scientists had actually read his original 1968 paper, and even fewer really knew his argument based on boundary conditions. What everyone understood was the workings and ramifications of the flat lens. And from here sprang a myriad of papers all assuming that a negative index of refraction was a reality even if it had never been shown conclusively in this author's opinion (too weak a signal), as discussed extensively in Chapter 1. What was put in place very quickly was a "theory" that in a seemingly logical way explained "everything." Very creative minds invented materials called double positive, double negative, and so on. It was assumed in general that all these features could be obtained by using periodic structures (a notion that made me "freak out"). Most scientists simply like to read a logical layout that "explains" everything. Most of them have spent the majority of their time studying true and well-founded theories. Rarely did they have to doubt Maxwell's equations, and if they did, they were probably wrong! When I discuss my views in scientific meetings, the first reaction is invariably: But how can so many people be wrong? It simply takes quite awhile for many scientists to get used to the idea that they have been reading science fiction, not science.

From an educational point of view we should probably do more to train our students to be skeptical. In that respect I have found it very useful to review papers with students. Actually, it is great fun. And talking about reviews, I would gladly sign my reviews and have them printed along with the paper. I think the quality (and courtesy!) of the reviews would improve exponentially! Of course, it should be followed up by academic credit and not, as it is now, just be an opportunity for the reviewer to show his or her ego from behind a curtain. Let us have an open and frank discussion.

6.10   HOW FAR HAVE WE GONE IN OUR SELF-DECEPTION?

I recently reviewed a paper where the authors concluded that the intrinsic impedance of a left-handed material was negative. To them it indicated that "something was wrong." Reviewer 1 thought otherwise. He quoted a well-known researcher (no names, due to the review process): "On sign and branch of certain parameters for simple, lossy double-negative materials, the sign of these parameters is a matter of choice, and that a positive as well as negative sign can be used, as long as it is used consistently in a given solution." However, on p. 14 in ref. 17, the following is stated:

$$"Z_i = \frac{E_i^+}{H_{z,i}^r} = \frac{\omega \mu_i}{\lambda_{x,i}} \qquad (6.6)$$

Thus, as both μ_2 and $\lambda_{x,2}$ are negative in the left-handed medium, the impedance Z_i *is positive as is required for a passive medium.*"

The difference between the two conclusions above is that the first is based on pure math while the second is rooted in the real, physical world. In other words, we see basically the same problem as discussed in Sections 1.10, 1.11, and 6.4 to 6.6, where we stated that whereas, for example, time can be negative in the mathematical world, this is not an option in the physical.

Says this author: Why waste more time on left-handed materials? They have never been found in nature, has never been produced artificially, and probably just cannot be made. Certainly, everyone is entitled to spend his or her time pondering whatever subject he or she chooses, such as: What if the mechanical mass could be negative? Certainly, such a material has not yet been found on this planet, but maybe it could be located clinging to the ceiling in a cave. It would have a great future for producing "antigravity pills." Sure, the Air Force would be interested. Frankly, it has as much chance of being realized as has the flat lens that relies on negative time (see Sections 6.4 and 6.5).

6.11   BUT DIDN'T ANYONE SUSPECT ANYTHING?

Not long ago I asked a leading pioneer in this area if he still believed in a negative index of refraction. His answer: "Well, I have so many contracts. It depends on where I go. Some places I believe it is possible, other places I don't." When I told another scientist about my conclusions, he merely said, "Well, fine. But just don't talk about it."

Were they concerned about their funding, which has been flowing so freely? Frankly, I don't think we have created an environment that has promoted good science. Science fiction, yes, but not science. Of course, both of these scientists had a very impressive number of publications. However, nobody should be rewarded for just writing large numbers of papers. It should be the significance of the research itself that carries the day. Too often these two tasks are on a collision course.

Frankly, it seems strange that very few of the highly educated researchers in this field suspected that "all was not well." And if they did, why did they not speak up? Without insinuating anything, is truth not more important than funding? Also, why are any references to the theory of periodic structures left out almost systematically? Many of the standard misconceptions could have been weeded out just by reading, for example, refs. 15 and 16.

6.12   HOW REALISTIC ARE SMALL ARRAYS?

In this day and age, when "real estate" on a "platform" is at a premuim, it has long been of interest to reduce the size of antennas in general, provided that the gain and bandwidth remain the same. It has quite often been suggested that the use of metamaterials of some sort would make such a dream come true. Let us remind the reader about a few facts that should be well known to competent antenna scientists but apparently are very often overlooked.

The *directivity* of an antenna depends only on the size of the aperture measured in wavelength in air (provided that we transmit into free space) as well as the shape of the aperture distribution. It is immaterial whether the array elements are fed directly from individual generators or arrive via an exotic material. For a given-size aperture the directivity depends only on the shape of the aperture distribution: for example, a uniform distribution leads to maximum directivity, while lower values are obtained for tapered distribution (but also lower sidelobes). If the spacings between the elements are smaller than about $\lambda/2$ (depending on scan), adding more elements will not change the directivity provided that the aperture distribution is essentially unchanged. However, the *impedance* of the array will, in general, be more broadbanded by making the interelement spacing smaller (see Chapter 6 in ref. 15).

No computer program, no matter how exotic, can ever lead to new designs that violate these fundamental concepts. It is deplorable that so much time has been spent in front of screens where the operator may have perceived reality but very often was just playing Russian roulette!

NOTE: As this book is almost into final printing, the author became aware of a most recent publication [21]. This paper is of great interest because it not only agrees with our findings but uses an entirely different approach, namely, thermodynamic. It shows that for negative index materials, "negative heat" would be developed. In this book we conclude that "negative time" is essential in Veselago's medium. Obviously, both of these requirements are physically unrealistic.

REFERENCES

[1] V. G. Veselago, The electrodynamics of substance with simultaneously negative values of ε and μ, *Sov. Phys. Usp.*, vol. 10, pp. 509–524, Jan. 1968.

[2] J. B. Pendry, A. J. Holden, W. J. Stewart, and I. Youngs, Extremely low frequency plasmons in metallic mesostructure, *Phys. Rev. Lett.*, vol. 76, pp. 4773–3776, June 1996.

[3] J. B. Pendry, A. J. Holden, D. J. Robbins, and W. J. Stewart, Low frequency plasmons in thin-wire structures, *J. Phys. Condens. Matter*, vol. 10, pp. 4785–4809, 1998.

[4] J. B. Pendry, A. J. Holden, D. J. Robbins, and W. J. Stewart, Magnetism from conductors and enhanced nonlinear phenomena, *IEEE Trans. Microwave Theory Tech.*, vol. 47, pp. 2075–2084, Nov. 1999.

[5] J. B. Pendry, Negative refraction makes a perfect lens, *Phys. Rev. Lett.*, vol. 85, pp. 3966–3969, Oct. 2000.

[6] D. R. Smith, W. J. Padilla, D. C. Vier, S. C. Nemat-Nasser, and S. Schultz, Composite medium with simultaneously negative permeability and permittivity, *Phys. Rev. Lett.*, vol. 84, pp. 4184–4187, May 2000.

[7] D. R. Smith, D. C. Vier, N. Kroll, and S. Schultz, Direct calculation of permeability and permittivity for left-handed metamaterials, *Appl. Phys. Lett.*, vol. 77, pp. 2246–2248, Oct. 2000.

[8] R. A. Shelby, D. R. Smith, S. C. Nemat-Nasser, and S. Schultz, Microwave transmission through two-dimensional, isotropic, left-handed metamaterials, *Appl. Phys. Lett.* vol. 788, pp. 489–491, Jan. 2001.

[9] R. A. Shelby, D. R. Smith, and S. Schultz, Experimental verification of a negative refractive indes of refraction, *Science*, vol. 292, pp. 77–79, Apr. 2001.

[10] N. Engheta, Compact cavity resonators using metamaterials with negative permittivity and permeability, *Proceedings on Electromagnetics in Advanced Applications (ICEAA01)*, Torino, Italy, 2001.

[11] N. Engheta, Is Foster's reactance theorem satisfied in double-negative and single-negative media?, *Microwave Opt. Tech. Lett.*, vol. 39, pp. 11–14, Oct. 2003.

[12] P. M. Valanju, R. M. Walser, and A. P. Valanju, Wave refraction in negative-index media: always positive and very inhomogeneous, *Phys. Rev. Lett.*, vol. 88, May 2002.

[13] S. Ramo, J. R. Whinnery, and T. VanDuzer, *Fields and Waves in Communication Electronics*, 3rd ed., Wiley, New York, 1994.

[14] J. A. Kong, *Electromagnetic Wave Theory*, 2nd ed., EMW Pub., Cambridge, MA, 2000.

[15] B. A. Munk, *Finite Antenna Arrays and FSS*, Wiley, Hoboken, NJ, 2003.

[16] B. A. Munk, *Frequency Selective Surfaces: Theory and Design,* Wiley, New York, 2000.

[17] R. Marques, F. Matrin, and M. Sorolla, *Metamaterials with Negative Parameters: Theory, Design, and Microwave Applications*, Wiley, Hoboken, NJ, 2008.

[18] N. Engheta and R. W. Ziolkowski, *Metamaterials: Physics and Engineering Explorations*, Wiley, Hoboken, NJ, 2006.

[19] B. V. Eleftheriades and K. G. Balmain, *Negative-Refraction Metamaterials, Transmission Line Theory and Microwave Applications*, Wiley, Hoboken, NJ, 2005.

[20] C. Caloz and T. Itoh, *Electromagnetic Materials, Transmission Line Theory and Microwave Applications*, Wiley, Hoboken, NJ, 2006.

[21] Vadim A. Markel, Correct definition of the Poynting vector in electrically and magnetically polarizable medium reveals that negative refraction is impossible, *Optics Express*, vol. 23, 2008.

APPENDIX A
The Paper Rejected in 2003

A.1 COMMENTS WRITTEN IN 2007 CONCERNING MY REJECTED PAPER SUBMITTED IN 2003

This author always had doubt about the existence of a negative index of refraction. I alluded to that at the ICEAA01 conference in Torino, Italy, in 2001 when I pointed out from the floor that the results presented by Engheta [1] violated Foster's reactance theorem (my results for μ, $\varepsilon < 0$ would have rotated counterclockwise in the Smith chart in Figure 1.2). However, not before 2003 did I submit for publication in *IEEE Transactions on Antennas and Propagation* a paper with the inflammatory title: "On Negative μ_1 and ε_1: Fact and Fiction." Here I analyzed, without actual calculations, scattering from a periodic structure comprised of a single layer of infinitely long wires and double split-ring resonators as it would have been done by a person who had no knowledge (or interest) in materials with μ, $\varepsilon < 0$. As is well known, it leads to an equivalent circuit consisting of an infinite transmission line with a shunt impedance. If the structure is lossless, the shunt impedance is purely reactive (i.e., it will be located on the rim of the Smith chart and rotate clockwise as the frequency increases; see Figure 2 in Section A.2. However, if the structure is lossy, the shunt impedance will be lossy as well and be located at least partially inside the Smith chart, where Foster's reactance theorem no longer holds. In fact, when observed from the center of the Smith chart, the shunt impedance can indeed run counterclockwise over a limited frequency range (see, e.g., Figure 4 in Section A.2). Loss in the shunt impedance can have various causes. It may simply be due to the usual ohmic and dielectric losses in the structure itself. Such losses will be particularly noteworthy if two resonances are close

Metamaterials: Critique and Alternatives, By Ben A. Munk
Copyright © 2009 John Wiley & Sons, Inc.

to each other, as, for example, in a double split-ring resonator (again see Figure 4). Or it could be due to (God forbid!) the presence of grating lobes. These will supply real energy in the grating lobe directions and manifest themselves as resistive parts of the shunt impedance. Such a situation may indeed lead to transmission curves running counterclockwise over a limited frequency range. In fact, a recent paper [2] claimed experimental evidence of "left-handed" material consisting merely of conducting plates with circular holes. Surely enough, the distance between the holes was about λ, which would result in grating lobes even for a normal angle of incidence. The phase of the field transmitted can indeed run "backward" in such a case; however, $\bar{E}, \bar{H}$, and direction of propagation in any direction will always form a "right-handed" triplet (see Section 1.6). There is absolutely nothing "left-handed" or backward traveling about the transmitted field coming from such a structure. Besides, grating lobes would weaken the field in the principal direction of propagation. You can forget about the perfect lens when using this design.

As far as the actual rejection is concerned, there were a total of three reviewers. One found it interesting (although at times a bit polemic!), while the other two were adamantly against it. One of them even stated: "This paper should never be published, not even in the magazine." (Sorry, Ross, despite all the wonderful articles you have brought us over the years, this is what he said!)

Since my next promotion hardly depends on publication of a paper or two (after all, I am long retired), I decided not to waste my time fighting a couple of incompetent reviewers (they were simply a gold mine for one of my favored entertainments: common misconceptions). I believe the paper was important in 2003 and even more so today, five years later. So here it is for readers to judge for themselves. You may find the language a bit "juicy" at times. For example, the two "negative" reviewers were somewhat upset by my remarks at the end of Section 6.

REFERENCES

[1] N. Engheta, Compact cavity resonators using metamaterials with negative permittivity and permeability, *Proceedings on Electromagnetics in Advanced Applications (ICEAA01)*, Torino, Italy, 2001.

[2] M. Beruete, I. Campillo, M. Navarro-Cía, F. Falcone, and M. Sorolla Ayza, Molding left- or right-handed metamaterials by stacked cutoff metallic hole arrays, *IEEE Trans. Antennas Propag.*, vol. 55, pp. 1514–1521, June 2007.

A.2   THE PAPER REJECTED IN 2003

On Negative μ_1 and ε_1: Fact and Fiction

Ben A. Munk

Abstract

Materials perceived as having both negative μ_1 and ε_1 have been of considerable interest for some time. Apart from purely theoretical interest, it holds out the possibility for producing lenses with super-resolution to mention just a single feature [1]. However, this paper is not about possible applications but simply about the very existence of such materials as well as interpretation of various experiments. We conclude that "negative refractions" seems inconsistent with basic electromagnetic theory and that the observed scattered field from models of these metamaterials are most likely caused by radiation from certain types of surface waves unique to finite periodic structures with inter-element spacings less than $\lambda/2$.

1. Background: Do we really need μ_1 and ε_1?

Using the constitutive relationships

$$\bar{B} = \mu_1 \bar{H} \tag{1}$$

$$\bar{D} = \varepsilon_1 \bar{E} \tag{2}$$

we can write Maxwell's curl equations:

$$\nabla \times \bar{E} = -j\omega\bar{B} = -j\omega\mu_1\bar{H} \tag{3}$$

$$\nabla \times H = j\omega\bar{D} + \bar{J} = j\omega\varepsilon_1\bar{E} + \bar{J} \tag{4}$$

Further, we have for the intrinsic impedance

$$Z_1 = \sqrt{\frac{\mu_1}{\varepsilon_1}} \tag{5}$$

and the propagation constant

$$\beta_1 = \omega\sqrt{\mu_1\varepsilon_1} \tag{6}$$

Substituting (5) and (6) into (3) and (4) yields for a source free region $(\bar{J}=0)$:

$$\nabla \times \bar{E} = -j\beta_1 Z_1 \bar{H} \tag{7}$$

$$Z_1 \nabla \times \bar{H} = j\beta_1 \bar{E} \tag{8}$$

The first set of Maxwell's equations as given by (3) and (4) are seen to contain μ_1 and ε_1 directly. Thus, if we change signs of μ_1 and ε_1 it opens up the possibility for all kinds of sign changes for $\bar{E}$ and $\bar{H}$ as well. Or does it really? In fact, if we instead consider the second set of Maxwell's equations as given by (7) and (8), we observe that they contain only the intrinsic impedance Z_1 and the propagation constant β_1. Further, we see from (5) and (6) that if we change sign of both μ_1 and ε_1, no change will be observed for either Z_1 or β_1, and consequently not of the field vectors $\bar{E}$ and $\bar{H}$, either. That is, unless we choose the negative sign associated with the square root for β_1 as given by (6). We will discuss this choice later several times.

It should be pointed out that Veselago in his original 1968 paper [4] is fully aware that expressions containing products of μ_1 and ε_1 are invariant when we change signs for both μ_1 and ε_1. He further points out that only when μ_1 or ε_1 appears alone, as in (3) and (4) above, can we expect to gain insight into new materials. Further, $\bar{E}$, $\bar{H}$ and $\bar{\beta}_1$ form a right-handed triplet of vectors when $\mu_1 > 0$ and $\varepsilon_1 > 0$, while Veselago states that they form a left-handed system if μ_1 and ε_1 are negative. He made no claims whether such left-handed materials actually exist but speculated that they might be realized in plasmas under certain circumstances. Note, that choosing the negative sign for β_1 apparently brings us in conceptual agreement with Veselago's left-handed system. However, we will not go into this subject in this paper but will limit our investigation to the widely publicized artificial metamaterials that consist merely of periodic structures made of various wire elements suspended on a dielectric substrate. Note that these imitations of left-handed materials are made entirely of right-handed materials. Consequently, $\bar{E}$, $\bar{H}$ and $\bar{\beta}_1$ do not change handedness when entering the artificial material and its properties can be calculated in the conventional way.

But why are we customarily using (3) and (4), rather than (7) and (8)? Primarily because of tradition. We simply inherited μ_1 and ε_1 from static electromagnetism where they can be obtained individually by direct measurements. In the dynamic range we usually measure Z_1 and β_1 and from there we find μ_1 and ε_1 by application of (5) and (6). It is somewhat ironic that when a scientist learns about the values of μ_1 and ε_1 for a new material, he quickly determines Z_1 and β_1. That will indicate what the typical use could be for that new material as, for example, a low loss dielectric for a radome or for an absorber. In other words, are we not fussing a little bit too much over μ_1 and ε_1? And thereby perhaps ending up on the wrong street?

2. But are artificial materials with negative μ_1 and ε_1 not a reality? Not necessarily!

Certainly, artificial materials have been produced that claim large negative values of both μ_1 and ε_1. They are usually made of a periodic structure with elements typically shaped similar to a "horseshoe" [3] as shown in Figure 1a. Interlaced end-loaded dipoles or just straight wires are sometimes also added [3–6]. A typical equivalent circuit is shown in Figure 1b. The horseshoe elements are merely forming a parallel combination of coils and capacitors with a resonant frequency, f_2, observed when the circumference of the horseshoe is $\sim\lambda/2$ (in the dielectric substrate, if any). Further, there will be a significant capacitance C_e between adjacent elements.

This will result in a series resonance frequency, $f_1 < f_2$, when the parallel circuit is inductive. The equivalent circuit in Figure 1b may be summarized in the equivalent circuit of Figure 1c where we show a transmission line with characteristic impedance Z_1 and a shunt reactance jX_S. It is very instructive to plot jX_S in a Smith chart as shown in Figure 2 [7,8]. Since it is purely reactive (for no ohmic loss) it will be located at the rim of the Smith chart. At the DC frequency, $f_0 = 0$, $jX_S = \infty$, which is also the case at f_2 as shown. The reflection coefficient at these two frequencies will be zero as indicated in Figure 2b. However, at the frequency f_1, $jX_S = 0$ resulting in 100% reflection as also shown. (See

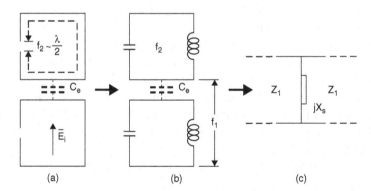

(a)   (b)   (c)

Figure 1   An artificial dielectric with perceived negative μ_1 and ε_1 is typically made as a periodic structure with various compact elements. (a) The most common element used for metamaterials with μ_1, $\varepsilon_1 < 0$, namely the horseshoe element. (b) Equivalent circuit of horseshoe element shown in (a). It exhibits an antiresonance at f_2 where the circumference of the horseshoe is $\sim\lambda/2$ (in dielectric, if any). Further, due to the capacitance, C_e, between adjacent elements, it exhibits a resonance at $f_1 < f_2$. (c) The reactance observed in (b) can be summarized as a shunt reactance jX_s on an infinite transmission line with characteristic impedance, Z_1.

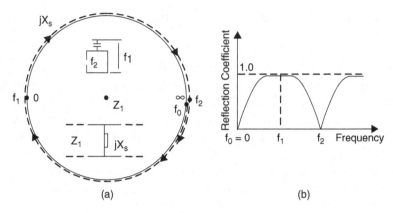

(a)   (b)

Figure 2 (a) The shunt reactance jX_s shown in the insert is purely reactive and therefore located on the rim of the Smith chart. It is infinite at $f_0 = 0$ and f_2 while it is a short circuit at f_1. (b) The reflection coefficient as seen by an incident wave from the left side in the insert when the right side is terminated in Z_1 (or the cable infinite long).

also [8], chapter 9.) Note that precise calculations based on, for example, the PMM program will verify this representation that is based primarily on physical insight valid for small elements.

As we move up in frequency we may encounter more resonances and anti-resonances alternating among each other according to Foster's Reactance Theorem, i.e. moving clockwise. That is, until we reach onset of grating lobes, in which case the reactance jX_S becomes lossy due to radiation in the grating lobe direction(s) and it moves inside the Smith chart. We can accelerate this rotation in the Smith chart by adding more resonating elements for example in the form of smaller horseshoe elements inside the original ones as shown in the insert of Figure 3a. We show here a Smith chart where we note that new resonant and antiresonant frequencies, f_3 and f_4, respectively, have been added as a result of the inside horseshoe element. Again, as long as there is no loss and no grating lobe, the shunt impedance, jX_S, for small elements will be purely imaginary and located entirely on the rim. The reflection coefficient is shown in Figure 3b.

A very instructive experiment is now to move the frequencies f_2 and f_4 close to each other for example by making the inside element only slightly smaller than the outside. Further, it is in that case realistic to add losses to our model, either ohmic and/or dielectric. In that event the shunt reactance becomes lossy and will move away from the rim to the inside of the Smith chart as shown in Figure 4a. In particular the frequency point f_3 can in extreme cases move significantly to the right as shown.

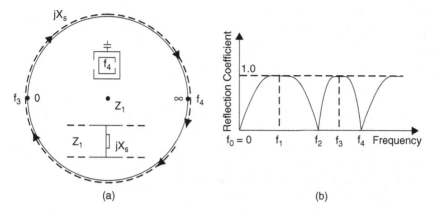

Figure 3   (a) By placing a smaller horseshoe element inside the longer one in Figure 2(a), we can introduce the higher resonance frequencies f_3 and f_4. (b) The reflection coefficient as seen by a wave incident from the left side in the insert when the right side is terminated in Z_1.

The reflection coefficient is shown in Figure 4b. Note that the curve is actually running counter clockwise in the neighborhood of f_3. This in no way violates Foster's reactance theorem that pertains only to lossless systems [9]. The "backward rotation" of the curve is sometimes interpreted as the existence of a material with negative μ_1 and ε_1. However, if we normalize our Smith chart to a point inside the f_3 loop, the curve will run clockwise and show no "abnormal" behavior. Still, our "composite

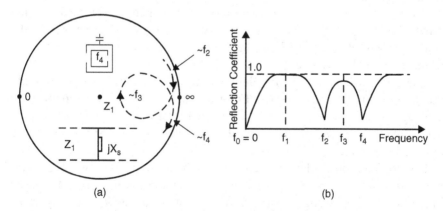

Figure 4   (a) When moving the two "horseshoe" frequencies, f_2 and f_4, closer to each other, the impedance at f_3, under the influence of loss, will move closer to f_2 and f_4 as shown. (b) The reflection coefficient as seen by an incident wave from the left side in the insert when the right side is terminated in Z_1.

material" is the same. This confusion seems to be related to splitting the complex curve into real and imaginary components. Bad idea unless done right!

3. But what about the index of refraction, n_1, and propagation constant, β_1?

The intrinsic impedance Z_1 and the propagation constant β_1 are given by (5) and (6), respectively. As already pointed out, changing the signs of μ_1 and ε_1 simultaneously does not change either Z_1 or β_1, but we may choose the negative sign for β_1.

Further, we still have to discuss the index of refraction as given by

$$n_1 = \sqrt{\frac{\mu_1 \varepsilon_1}{\mu_0 \varepsilon_0}} \tag{9}$$

Obviously, if we change the signs of both μ_1 and ε_1, we will not observe any change in n_1 either. However, mathematically speaking, the complete solution for n_1 is

$$n_1 = \pm\sqrt{\frac{\mu_1 \varepsilon_1}{\mu_0 \varepsilon_0}} \tag{10}$$

Here the $+$ sign simply bends an incident plane wave according to the well known Snell's Law. The negative value on the other hand will produce a wave propagating with a negative angle of refraction. It is believed by many that this phenomenon can be observed for materials with negative artificial μ_1 and ε_1 [1,3–6].

This problem has been theoretically investigated by Ziolkowski and others [10,11] but they have apparently not convinced everyone in the scientific community [12,13]. In fact, much more attention has been given to experimental evidence of the negative sign of n_1.*

The most widely performed demonstration of negative n_1 is probably the wedge experiment shown in Figure 5. A plane wave is incident upon a wedge with wedge angle, v, and index of refraction $n_1 > 0$ or $n_1 < 0$. The directions of propagation for the refracted waves for both of these cases are also shown.

Indeed, attempts to design such a wedge have been done by using meta-materials made of horseshoe elements as shown in Figure 1 and discussed in section 2. The wedge shape comes about by using various layers of arrays of unequal length as shown in Figure 6. Many research groups have

---

*It appears that few believe calculations except the one who did them. Everybody believes measurements except the guy who took them!

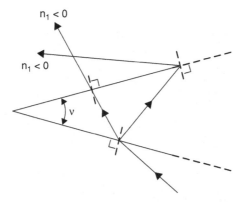

Figure 5 A wedge with wedge angle v is exposed to an incident field from below. We show the refracted field for the index of refraction $n_1 > 0$ (ordinary) as well as for $n_1 < 0$ (extraordinary).

reported a signal to the "negative side" that eventually could indicate the possibility of a negative n_1 [14,15]. However, probably the most interesting of these measurements were performed by a group from Boeing [16]. They noted that earlier measurements had been criticized because the refractions pattern was not measured in the far field. Consequently they performed measurement not only in the near field where they obtained pattern similar to Figure 6 alright, but also took measurements in the far field where the pattern looked somewhat different, namely as shown in Figure 7. It can best be described as being of similar shape as the near field pattern also shown in Figure 7 but intersected with several deep nulls. This agrees perfectly with the experience of all seasoned antenna

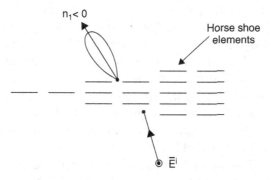

Figure 6 An artificial wedge made of horseshoe elements as shown in Figure 1. An incident signal $\bar{E}^i$ from below may produce a field on the top that may be interpreted as the artificial wedge has an index of refraction $n_1 < 0$.

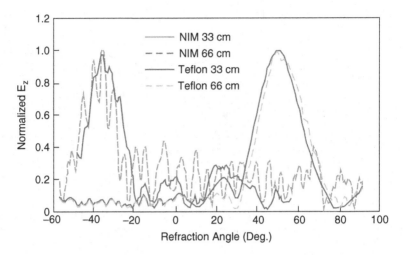

Figure 7 Experimental curves, as obtained by [16], of the refracted field for: (left) An artificial wedge of horseshoe elements, in the near field (33 cm) as well as the far field (66 cm). The negative angular location could be interpreted as a negative index of refraction. (right) The ordinary observed refracted field with $n_1 > 0$ for a Teflon wedge in the near field (33 cm) as well as the far field (66 cm).

engineers, namely, that as we move closer to a radiating structure, the first to go from the far field pattern are the nulls. The logical question is now: Is there any configuration that will produce a far field scattering pattern similar to the far field pattern shown in Figure 7? There is indeed. In fact, if we expose a finite array of straight wires to an incident plane wave it has been shown that a left and right going "surface" wave can be excited and radiate a pattern as shown in Figure 8. (See [8], Figure 4.9.) Typically this effect for straight dipoles is strongest at 20–30% below the resonance of the straight wires and we must require the inter-element spacing in the x-direction to be less than $\lambda/2$. This spacing requirement is indeed fulfilled for the horseshoe array in Figure 1 and although no calculations of such finite arrays of horseshoes are available at this time, we have little doubt that they, too, are able to carry surface waves with a radiation pattern similar to the straight wire case in Figure 8. It is further interesting to note that the magnitude of both the measured curves in Figure 7 and the calculated curves in Figure 8 are about 20 dB or more below the incident field. This is definitely not in agreement with calculations by Ziolkowski that claims calculated magnitudes as strong as ∼0.3 dB below the incident field [10]. Actually the deep nulls in the refraction pattern were also observed by Shelby et al. [14]. Only, they interpreted this as a flaw in their artificial wedge due to the element size

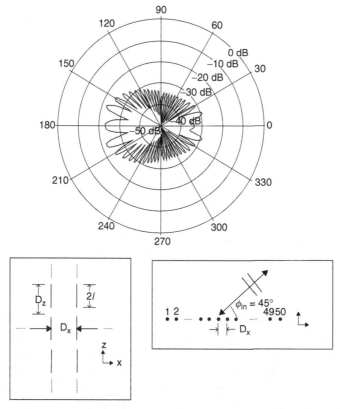

Figure 8 The calculated bistatic scattered field at $f = 7.7$ GHz from a finite × infinite array of 50 columns of straight dipoles resonating ∼10 GHz. Incident field as shown in insert. (From [8], Figure 4.9).

being too large. Consequently they averaged their measured results and thereby missed probably the most interesting feature of their experiment.

4. Further: No problems with boundary conditions when surface waves are introduced.

Figure 9 shows a plane wave in medium ① with direction of propagation, $\hat{s}_1$, entering medium ⓪ where the direction is denoted $\hat{s}_0$. Thus, the phase velocity along the boundary is $\beta_1 s_{1x}$ and $\beta_0 s_{0x}$ for medium ① and ⓪, respectively. Matching these two phase velocities yields $\beta_1 s_{1x} = \beta_0 s_{0x}$ and by noting that $s_{1x} = \sin\theta_1$ and $s_{0x} = \sin\theta_0$ we obtain Snell's Law

$$\frac{\sin\theta_1}{\sin\theta_0} = \frac{\beta_0}{\beta_1} = \sqrt{\frac{\mu_0\varepsilon_0}{\mu_1\varepsilon_1}} = \frac{1}{n_1} \tag{11}$$

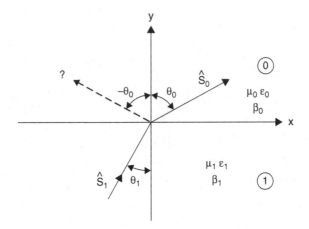

Figure 9 An incident wave medium ① with angle of incidence θ is being refracted in the direction θ_0 in medium ⓪. For a negative index of refraction, the refracted angle would be $-\theta_0$.

If $n_1 < 0$ as suggested by (10), we obtain sin $\theta_0 < 0$, i.e., a direction of propagation in media ⓪ as indicated by the broken line arrow in Figure 9. Obviously the phase velocities of the incident and the negative refracted wave at the boundary are physically incompatible. However, many proposals have been suggested to overcome this "minor" problem. One suggestion was simply to reverse the direction of the broken line arrow. [Why not? This whole subject is basically about signs anyway!] Another simply throws the hands up into the air and cries "Anomaly" (Indeed!). However, do not forget that our artificial material is located entirely in a right-handed and not a left-handed space. The fact is that only surface waves attached to some structure at the boundary can carry waves to the left (and right) without having to match the incident field. Such surface waves have phase velocities completely different from $\beta_1 s_{1x} = \beta_0 s_{0x}$, i.e., the field of this wave is completely different from the incident field. More specifically, we show in Figure 10 (upper) the actual calculated element currents in an array of 25 infinitely long columns of dipoles, each 1.5 cm long. Also shown in the same figure (lower) is the Fourier analysis of the actual currents. Note that we also obtain, in addition to the Floquet current at $r_{cx} = 0.707$, two opposite traveling surface waves at $r_{cx} = \pm 1.25$. Obviously, this phase velocity can never be matched to the incident wave where $|r_{cx}| < 1$. Note, however, that it is the incident wave that excites the surface wave that in turn radiates a typical pattern as shown in Figures 7 and 8. When the angle of incidence θ_1 is changed, the multi-lobed pattern will change somewhat alright but not according to

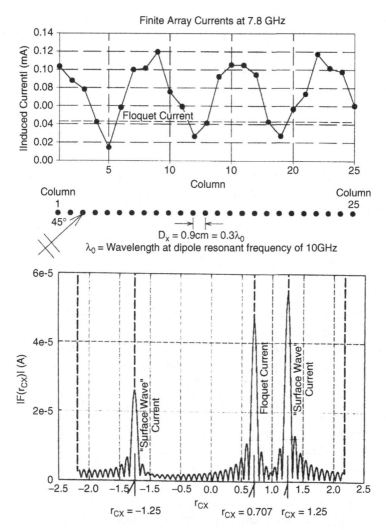

Figure 10   Top: Actual calculated element currents in an array of 25 infinite long columns of dipoles each 1.5 cm long. Angle of incidence is 45° as shown in insert. Bottom: A Fourier analysis of the actual calculated element currents above. Note that in addition to the Floquet current at $r_{cx} = 0.707$, we also observe two opposite traveling surface waves at $r_{cx} = \pm 1.25$. Also the surface wave at $r_{cx} = 1.25$ is considerably stronger than the Floquet current. However, it has less efficient radiation as observed in Figure 8.

the negative Snell's Law given by (10). For further details, see reference [8], chapter 4, and [17–20].

But how do we know that the surface wave is attached to the periodic structure and not just the dielectric interface? From the bandwidth. The periodic structure of straight dipoles can support strong surface waves over $\sim 10\%$ bandwidth [8] while purely dielectric surface waves would be more broadbanded. Finally, a periodic structure of horseshoe shape would be very narrow banded as has been lamented repeatedly by everybody. The fact is that the phenomenon attributed to the horseshoe element can be observed just as well or even better with a host of other elements.

5. Negative artificial propagation constant β_1 is an illusion: a direct proof.

The wedge experiment described above has apparently convinced many scientists that negative index of refraction is possible by using artificial metamaterials. Hopefully, the discussion here has indicated that this may not be quite the case. However, many participants in this endeavor would like to see an alternative and more direct proof. Thus, in the following example the propagation constant, β_1, will be investigated rather than the index of refraction, n_1. Keep in mind, however, that the sign for these two are identical as seen by inspection of (10) and the complete solution of (6). This proof is so direct and simple that one wonders, why it has not been presented before.

In Figure 11 we show an ordinary dielectric slab of thickness l_2 and propagation constant β_2. To the left we have placed a groundplane and to the right a slab of metamaterials with equivalent thickness l_1 and propagation constant β_1. Typically this is believed to come about by using one or more of the horseshoe arrays shown in Figures 1–3 and 4. We are going to determine the input impedance Z_{in1} at the right surface of the metamaterials slab l_1 looking left as indicated in Figure 11.

We will obtain Z_{in1} in two ways: (a) By using the Smith chart shown in Figure 12 and assuming $\beta_1 < 0$ as postulated by some scientists. (b) By direct calculation of the groundplane with periodic structures placed in front of it as shown schematically in the inset of Figure 11.

In Figure 12 we show how the input impedance Z_{in2} of slab l_2 is obtained by rotating the amount $\beta_2 l_2$ clockwise in the Smith chart from the groundplane (zero). Next we obtain Z_{in1} by rotating the amount $\beta_1 l_1$ in the counter clockwise direction from Z_{in2} (if the intrinsic impedances for slab l_1 and l_2 differ, it is a simple matter to adjust for this in the Smith chart). The important point is not so much where we end up on the rim of the Smith chart as the fact that as the frequency increases, Z_{in2} rotates clockwise and the contribution from slab l_1 moves opposite. Obviously, if $l_2 \sim 0$, Z_{in1} will always move counter clockwise for increasing frequency because of the assumed negative value of β_1.

Alternatively, we can also obtain Z_{in1} by direct calculation; for example, by using the PMM program. However, for the purpose of

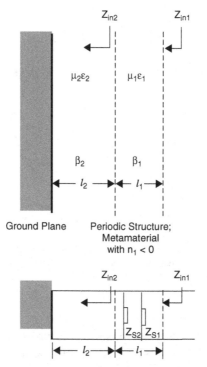

Figure 11   Top: A groundplane followed by a right-handed slab with material constants $\mu_2\varepsilon_2$ and thickness l_2. Next follows a section of one or more wire FSS arrays. Their total perceived thickness is l_1 and the material constants $\mu_1\varepsilon_1$. Bottom: The approximate equivalent circuit when the number of FSS arrays equals two.

determining the direction of Z_{in1} for increasing frequency, it is perfectly adequate to use the equivalent circuit for each FSS sheet as given in Figures 1 and 2 and shown in Figure 11, lower part, for two sheets. Obviously, from Foster's reactance theorem [9] we can immediately conclude that Z_{in2} will rotate clockwise with increasing frequency regardless of now many FSS sheets we use. Thus, the propagation constant β_1 for the artificial metamaterials can never be negative as assumed above.

Further, as alluded to in the introduction, this example also shows that the values of μ_1 and ε_1 are only of secondary interest. It is Z_1 and β_1 that really "run the show."

6. Is there any other way to obtain negative β_1 and n_1?

At this point it appears that a negative propagation constant, β_1, as well as a negative index of refraction, n_1, cannot be obtained by use of periodic structures as we know them today. However, is it possible by

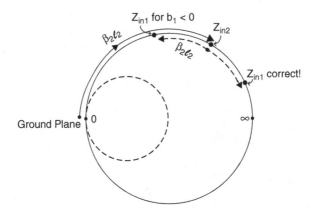

Figure 12   Calculation of the input impedance Z_{in1} two ways: First, by using the circuit in Figure 10, top, when we assume $\beta_1 < 0$. Second, by direct calculation using the PMM program by placing actual FSS arrays in front of slab l_2 as shown in Figure 10, bottom. Alternatively, the sign of β_1 is easily obtained by application of Foster's reactance theorem on the equivalent circuit.

any other techniques? Many scientists would probably suggest "Metamaterials." Now, this is a relatively new discipline and a precise definition seems not to have crystallized yet (at least not to this author's knowledge). However, as the mystique somehow evaporates, it consists essentially of a three dimensional periodic structure. This author sees no reason why these should behave radically different than the multilayered periodic surfaces considered above and also in reference [7], chapters 7 and 8. In other words, the emergence of negative β_1 and n_1 is not very likely. If anyone pulls it off, please be advised not to write a paper. Just see the author of this paper immediately.

7. Conclusions

We have discussed how μ_1 and ε_1 can actually be replaced by the intrinsic impedance $Z_1 = \sqrt{\mu_1/\varepsilon_1}$ and the propagation constant $\beta_1 = \omega \sqrt{\mu_1 \varepsilon_1}$. Obviously, when changing signs of both μ_1 and ε_1, no change is observed in Z_1 and β_1 and not in the field vectors unless we choose $\beta_1 < 0$. However, one is, of course, still entitled to ponder the question whether materials with negative μ_1 and ε_1 do exist at all. Typically they are attributed to periodic structures made of horseshoe shaped elements, simple endloaded dipoles or both. It was pointed out that such structures have an equivalent circuit consisting of a transmission line with a shunt reactance jX_s having alternative poles and zeros according to Foster's reactance theorem (until onset of grating lobes). However, this is typical for most periodic structures. In other words, if negative μ_1 and ε_1 can be extracted from horseshoe elements and dipoles, the same can be done with a myriad of other infinite periodic structures. We simply don't think this is possible unless loss is introduced as discussed above in section 2.

The question whether negative indices of refraction and propagation constant actually exist is probably the most controversial. First of all, we must realize that this problem has most likely very little to do with whether μ_1 and ε_1 are positive or negative but a lot to do with the sign we chose for $n_1 = \pm\sqrt{\mu_1\varepsilon_1/\mu_0\varepsilon_0}$. We discussed in detail that choosing the negative sign would lead to inconsistencies in the boundary conditions between the incident and refracted fields. However, we also pointed out that it is indeed possible to have both a left and a right going surface wave existing on a periodic structure. We emphasize that this type of surface wave can exist only on a *finite* periodic structure and only with inter-element spacing less then $\lambda/2$ [8]. Note further that phase velocities of the surface waves and the X-component of the incident field are completely different. Thus, we are not talking here about matching fields and therefore not about a negative index of refraction. In other words, when we change the angle of incidence the "refracted" field will not change like a negative Snell's Law. The field observed by measurement is simply the field radiated by surface waves unique to finite periodic structures. The calculated pattern of these surface waves and the measured refraction pattern looks very similar, namely multi-lobed, and their magnitudes are down ~ 20 dB or more from the incident field.

We finally demonstrated, by direct calculation using the equivalent circuit for FSS sheets, that the propagation constant β_1 (and thereby the index of refraction, n_1) is always positive.

It should finally be emphasized that measurements of dielectric substrates with horseshoe and other elements has been performed at numerous laboratories. To the best of our knowledge nobody has actually measured either negative μ_1 or ε_1. The result of their findings were that the samples were dielectric slabs with wires that sometimes could make the reflected field go backward in the Smith chart. However, such a feature is typically observed for broadband matching as discussed in [8]. It has nothing to do with negative μ_1 and ε_1. This effect is sometimes denoted the "Mambrino Effect."

It appears that we are right back to Veselago's paper from 1968. He never claimed that materials with real negative μ_1 and ε_1 actually existed on this planet. Do they in outer space? Are "black holes" related to such materials? Should we choose the negative sign for Z_1 in (5)?

REFERENCES

[1] J. B. Pendry, Negative refraction makes a perfect lens, *Phys. Rev. Lett.*, vol. 85, no. 18, pp. 3966–3969, Oct. 2000.

[2] V. G. Veselago, The electrodynamics of substances with simultaneously negative values of ε and μ, *Sov. Phys, Usp.*, vol. 10, no. 4, pp. 509–524, Jan.–Feb. 1968.

[3] R. W. Ziolkowski, Design, fabrication, and testing of double negative metamaterials, *IEEE Trans. Antennas Propag.*, vol. 51, no. 7, pp. 1516–1529, July 2003.

[4] D. R. Smith, W. J. Padilla, D. C. Vier, S. C. Neamt-Nasser, and S. Schultz, Composite medium with simultaneously negative permeability and permittivity, *Phys. Rev. Lett.*, vol. 84, no. 18, May 2000.

[5] D. R. Smith, D. C. Vier, N. Kroll, and S. Schultz, Direct calculation of permeabilitiy and permittivity for a left-handed metamaterial, *Appl. Phys. Lett.*, vol. 77, no. 14, Oct. 2000.

[6] R. A. Shelby, D. R. Smith, S. C. Nemat-Nasser, and S. Schultz, Microwave transmission through a two-dimensional, isotropic, left-handed metamaterials, *Appl. Phys. Lett.*, vol. 78, no. 4, Jan. 2001.

[7] B. A. Munk, *Frequency Selective Surfaces: Theory and Design*, Wiley, New York, 2000, App. A and Sec. 4.4.1.

[8] B. A. Munk, *Finite Antenna Arrays and FSS*, Wiley, Hoboken, NJ, 2003, Chap. 1 and Apps. A and B.

[9] R. M. Foster, A reactance theorem, *Bell Syst. Tech. J.*, vol. 30, pp. 259–267, Apr. 1924.

[10] Ziolkowski, op. cit. and private communication.

[11] D. R. Smith and N. Kroll, Negative refractive index in left-handed materials, *Phys. Rev. Lett.*, vol. 85, no. 14, Oct. 2000.

[12] P. M. Valanju, R. M. Walser, and A. P. Valanju, Wave refraction in negative-index media: always positive and very inhomogeneous, *Phys. Rev. Lett.*, vol. 88, no. 18, May 2002.

[13] N. Garcia and M. Nieto-Vesperinas, Left-handed materials do not make a perfect lens, *Phys. Rev. Lett.*, vol. 88, no. 20, May 2002.

[14] R. A. Shelby, D. R. Smith, and S. Schultz, Experimental verification of a negative refractive index of refraction, *Science*, vol. 292, pp. 77–79, Apr. 2001.

[15] A. Grbic and G. V. Eleftheriades, Experimental verification of backward-wave radiation from a negative refractive index metamaterials, *J. Appl. Phys.*, vol. 92, no. 10, pp. 5930–5935, Nov. 2002.

[16] C. G. Parazzoli, R. B. Greegor, K. Li, B. E. C. Koltenbah, and M. Tanielian, Experimental verification and simulation of negative index of refraction using Snell's law, *Phys. Rev. Lett.*, vol. 90, no. 10, Mar. 2003.

[17] J. B. Pryor, Suppression of surface waves on arrays of finite extent, M.Sc. thesis, Ohio State University, 2000.

[18] D. Janning, Surface waves in arrays of finite extent, Ph.D. dissertation, Ohio State University, 2000.

[19] B. A. Munk, D. S. Janning, J. B. Pryor, and R. J. Marhefka, Scattering from surface waves on finite FSS, *IEEE Trans. Antennas Propag.*, vol. 49, no. 12, pp. 1782–1793, Dec. 2001.

[20] D. S. Janning and B. A. Munk, Effect of surface waves on the current of truncated periodic arrays, *IEEE Trans. Antennas Propag.*, vol. 40, no. 9, pp. 1254–1265, Sept. 2002.

APPENDIX B
Cavity-Type Broadband Antenna with a Steerable Cardioid Pattern

An IFF antenna: Where we show a logical development of a design concept, which unfortunately so often is missing in this day and age. You cannot just fool around with "magic" materials. A design concept is necessary, as shown here.

B.1 INTRODUCTION

This antenna was intended to be used as an IFF antenna operating somewhere in the range 1.0 to 1.25 GHz. The objective was to produce a radiation pattern with a broad main beam in the front sector and low radiation in the back sector. Further, the pattern should be steerable in at least four directions in the horizontal plane. In other words, we were looking for a steerable antenna with a cardioid pattern. Such an antenna would obviously be very useful in reducing jamming incident in the back sector.

B.2 DESIGN 1

It is well known that placing two omnidirectional antennas a quarter-wavelength apart and feeding them $90°$ out of phase will produce a cardioid pattern, as illustrated in Figure B.1. Since the antenna should be flush mounted, we used four loop elements rather than monopoles placed inside a circular cavity, as illustrated in Figure B.2. We notice further that the loops have been provided with series capacitors for matching purposes by drilling holes lined with Teflon sleeves inside small brass blocks located at the $90°$ bend. In this way the impedance of each of the four loops

Metamaterials: Critique and Alternatives, By Ben A. Munk
Copyright © 2009 John Wiley & Sons, Inc.

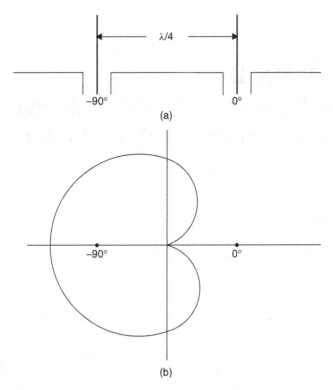

Figure B.1 The best-known way of producing a cardioid pattern: Two monopoles fed in quadrature and spaced λ/4 apart.

could be tuned to a value with VSWR < 2 if the other three loops were open-circuited. However, great changes in the single loop impedance were observed depending on the load condition of the other three. Thus, as we expected, there was strong mutual coupling among the four loops. Nevertheless, we managed to produce satisfactory cardioid patterns but only over a bandwidth of a few percent. In other words, this design was not satisfactory at all.

However, before we pursue alternative designs, we measured the impedance of each of the two active loops and determined the fundamental problem with this design: Whereas the impedance of the front loop was behaving "nicely" over the entire band, the impedance of the back loop was extremely bad: in fact, so bad that at some frequencies it wandered off the Smith chart! We see no reasons to present these results in detail here but refer interested readers to ref. 1 for details.

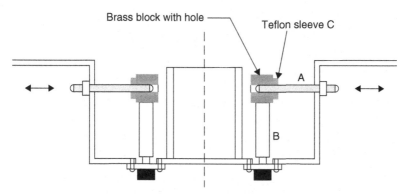

Figure B.2 The mechanical design of loops containing a series capacitance created between rod A and the hole in the brass block lined with Teflon sleeve C.

B.3   DESIGN 2

Although the results above revealed only a small portion of a rather long series of experimental work, we may quite safely conclude that the lack of pattern bandwidth can be traced back to a very strong mutual coupling between the loops in the cavity. It was further felt that the asymmetric nature of design 1 was primarily to be blamed for this calamity. Consequently, we concentrated on obtaining a more symmetrical design, as shown in Figure B.3. It consists of four loops similar to design 1. However, in addition, it has a top-loaded monopole mounted in the center. Further, any two opposing loops are always excited with currents of equal amplitude and exactly 180° out of phase. This assures that the coupling between the monopole and any two opposing pairs of loops will ideally be zero. The cardioid-shaped pattern is now created in the following way.

If we feed the four loops with equal amplitude but with a progressive phase (namely, 0°, 90°, 180°, and 270°), it can be shown that the radiation pattern for these four loops will essentially be circular in the horizontal plane provided that the average cavity diameter is less than 0.6λ at midband, as will, of course, the radiation pattern of the monopole in the middle. However, if we would record the phase of the far field of the monopole as we move around the antenna once at a certain distance, we would find it to be constant while the phase of the field from the four loops would change by 360°. Thus, at a certain direction in the horizontal plane the "loop field" and the monopole field will be exactly in phase, whereas in the opposite direction, they will be exactly out of phase. If, further, the

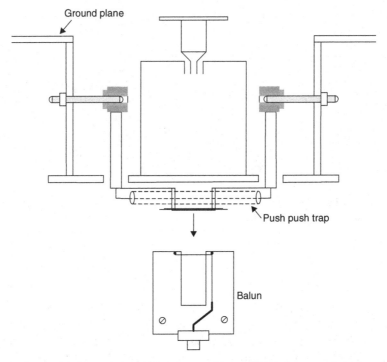

Figure B.3 Design 2, consisting of four loops with a top-loaded monopole in the middle. Feeding opposing loops in series from a balun assures the two loops to be exactly 180° out of phase.

two fields are of the same magnitude, a complete cancellation will take place in this direction.

A more complete understanding is perhaps best obtained by inspection of Figure B.4. Here we show the two figure-eight patterns from loop pairs $1+3$ and $2+4$. Since they are fed in quadrature, they will combine into a circular pattern coinciding with the monopole pattern in amplitude. The total pattern also shown in Figure B.4 can be obtained by simple use of phasors. It is also clear by inspection of Figure B.4 that loops $2+4$ not only "waste" all their radiated power out to the sides of the radiation pattern but also deliver no energy in the forward direction. Consequently, we might as well not excite loops 2 and 4, thus arrive at the modified pattern shown in Figure B.4b (design 2b). If we compare the two patterns in Figure B.4a and b, we note that curve a is down only 3 dB (with respect to ideal omnidirectional pattern at $\pm 90°$), while curve b is down 6 dB in the same directions. What is perhaps even more important is the fact that the null in the back direction is considerably broader for curve b than for

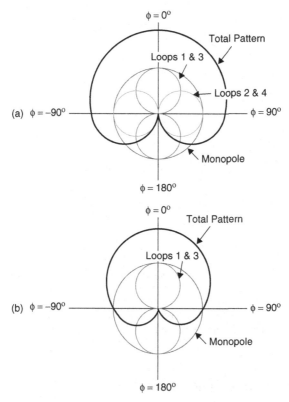

Figure B.4   (a) The cardioid pattern created by the monopole pattern and all four loops. (b) The cardioid pattern created by the monopole pattern and only two loops.

curve a. In fact, a more exact set of curves, shown in Figure B.5, bears out this statement very well. What's more, curve b has more directivity than a (estimated to be about 2 dB). On the other hand, the advantage by design 2a is that if we change the phase difference continuously between the group of four loops and the monopole, the null in the back direction will continuously rotate correspondingly in the horizontal plane, while design 2b is more suitable for stepwise rotation. However, when we consider that design 2b is lower than 15 dB for the back angular sector beyond $90°$ (see Figure B.5, curve b), this drawback becomes less important. Based on the foregoing considerations, we decided to actually develop design 2b, which would be somewhat simpler, lighter, and cheaper to build because we would not need a special 3-dB $90°$ hybrid to feed the two sets of loops. It should also be emphasized that design 2b could always be relatively easily modified into design 2a if it was later found to be desirable.

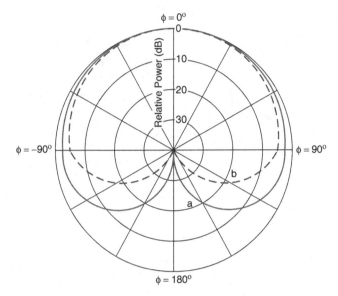

Figure B.5 Radiation patterns calculated for the two versions of design 2. Curve a: monopole and all four loops fed in phase rotation (design 2a). Curve b: Monopole and two opposing loops fed 180° out of phase (design 2b).

B.4 DEVELOPMENT OF DESIGN 2B

Design 2b can be considered to be made of two subantennas interlaced into each other: namely, one subantenna consisting of two loops producing the figure-eight pattern and the monopole producing the omnidirectional pattern. To produce a perfect figure-eight pattern, the two opposing loops are fed in series rather than in parallel. This will assure that the two loops are always fed 180° out of phase, as illustrated in Figure B.6a. For contrast, we also show in Figure B.6b the two loops fed in parallel, leading to in-phase feeding. (The terms *in phase* and *out of phase* refer here to a cylindrical coordinate system with vertical axis.)

Feeding the two loops 180° out of phase as in Figure B.6a requires the use of a balun. The first model is shown in Figure B.7a. It can be viewed as a stripline version of the coaxial version shown in Figure B.7b, used earlier by the author with excellent results (see Section 5.6). Basically, it consists of two transmission-line sections each of length about λ/4. They are connected at their far ends to the ground plane, where the unbalanced input is located. By connecting the inner conductor to the opposite outer conductor as shown, we obtain the balanced output in the middle of the construction. As mentioned above, the coaxial version of this balun has

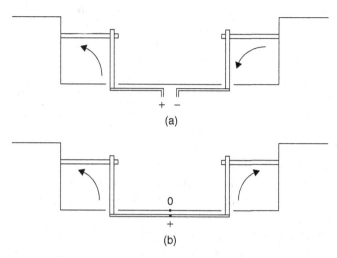

Figure B.6 (a) Two opposing loops being fed $180°$ out of phase by being fed in series (requires a balun). (b) Two opposing loops fed in-phase in parallel.

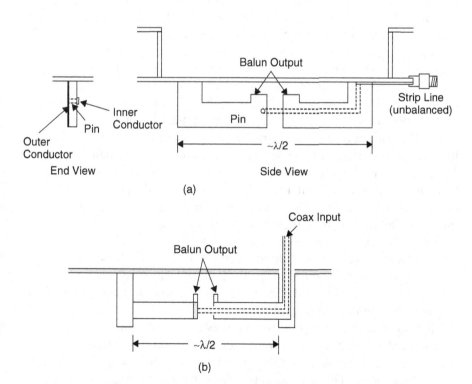

Figure B.7 (a) The first balun made in stripline. (b) The better-known coaxial forerunner for the stripline version above (see Section B.5).

been used before; however, the stripline version shown in Figure B.7a was less than successful; it radiated. In an attempt to eliminate this problem, we modified the stripline design into a microstrip design, but to no avail. Also, moving the two outer conductors closer to the ground plane showed only minor improvements, if any. We also thought that perhaps the unbalanced input leading from the balun to the right edge of the ground plane in Figure B.7a could somehow excite the entire ground plane, which consisted merely of slightly more than the cavity base (i.e., a 8×8 inch square). To this end we remounted the input connector at the edge of the ground plane as before, but this time in the plane orthogonal to the plane of the balun. Since this plane is electrically neutral, it was considered a better place to mount the input connector; however, again, this move was disappointing.

We finally decided that the ground plane available was too small for this type of balun to work satisfactorily, and we consequently decided to change to another type of balun. Although this new type ultimately proved successful, as we shall see below, we are in retrospect not quite sure that putting the blame on the size of the ground plane was correct. The fact is that the new balun also radiated independent of the ground plane but was subsequently brought under control, as explained later.

The new type of balun is shown in Figure B.8 and is merely a stripline version of the well-known Roberts balun. This type will provide a balanced output from dc up to frequencies when the transmission line deteriorates because of higher-order modes. Its bandwidth limitation in practice is determined by the fact that the length of the slit should be approximately an odd multiple of a quarter-wavelength in order not to affect the impedance properties of the antenna connected to the balanced output. This was no limitation in the present case (ca. 25%); in fact, the reactance from the slit in parallel with the antenna impedance at the balanced output can often be designed to make the combined "distorted" impedance more "broadbanded" than the original antenna impedance.

When this new balun was connected to a single pair of loops, we obtained a quite acceptable radiation pattern (i.e., nicely shaped symmetric figure eight with nulls more than 20 dB deep). However, when we next started taking patterns of the monopole in the middle (see Figure B.3), we ran into a multitude of problems.

First, the radiation pattern of the monopole was not omnidirectional but oval-shaped (at least at some frequencies). Second, the balun was "hot" (i.e., it radiated). The first problem was caused by the parasitic excitation of the two loops. The second was caused by the fact that the voltages induced in the two loops from the monopole were in phase rather than $180°$ out of phase. This phenomenon is illustrated in Figure B.9. We

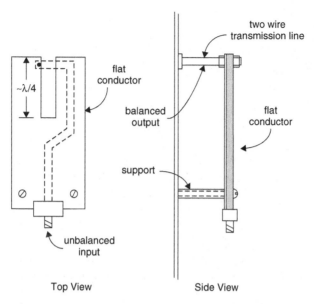

Figure B.8   The second balun is a stripline version of the well-known Roberts balun.

observe that while the push–push mode actually will never enter through the balun to the unbalanced input terminal, they will excite the entire balun in a push–push mode with respect to ground (i.e., the circuit board in Figure B.9).

The parasitic excitation of the loops leading to a distorted pattern of the monopole will of course depend very much on the load condition of the loops. It was determined experimentally that short circuiting the two loops close to these inputs to ground (i.e., at points A and A' in Figure B.9) produced the best monopole pattern. In other words, what was needed was a device that would short circuit the loops to ground when the voltages on the two loops produced push–push currents as shown in Figure B.9, while it would let the push–pull currents produced by a voltage at the loop input pass unobstructed.

B.4.1   Push–Push Traps

This action can possibly be accomplished in a number of ways. However, we believe that the simplest and most direct approach consists of merely connecting the two loops with a transmission line of length about $\lambda/2$. This is illustrated in Figure B.10 and works as follows: Imagine first that a push–push signal is applied to the "balanced" input (coming from the loops, or whatever), as illustrated in Figure B.11a. From each end

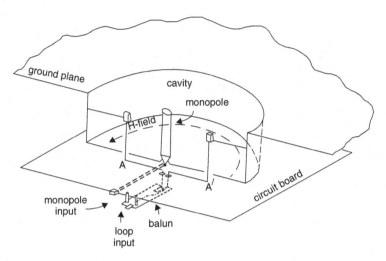

Figure B.9 The *H*-field from the monopole excites the two loops in phase, which subsequently makes the balun "hot" with respect to the ground plane (circuit board).

of the interconnecting cable, a wave will now travel toward the center where the two voltages will meet in phase (i.e., add together, just like the interconnecting cable was open-circuited at this point). Since the distance from this open circuit to the two ends is $\lambda/4$, we will obtain at the two endpoints the equivalent of a short circuit (i.e., the push–push mode on the transmission line will be stopped). On the other hand, if a push–pull mode is applied to the input as shown in Figure B.10b, the voltages meeting at the center of the transmission line will now be $180°$ out of phase (i.e., they will cancel, producing an equivalent short circuit at the center of the interconnecting cable). Consequently, an equivalent open circuit is produced at the two ends of the interconnecting cable, and the push–pull mode on the feeding lines will pass unobstructed.

In the practical execution of the push–push traps, we used RG-58U cable stripped of the PVC coating. Care should be taken to ensure good contact between the braids and the groundplane to prevent the trap from radiating.

B.4.2   Actual Layout

We are now ready to show the stripline layout of the antenna circuitry as given in Figure B.11. There are two inputs (to be combined later): one for the monopole and one for the loops. The latter is as explained earlier at the balun, and from here the balanced output is seen to be connected

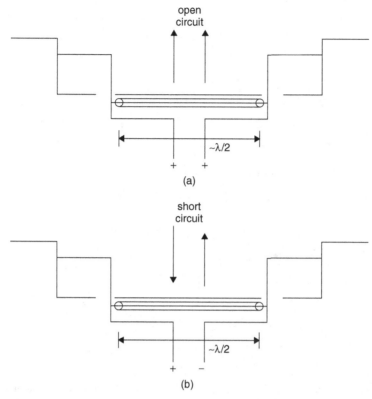

Figure B.10 The working of push–push traps. (a) A push–push signal will cause voltages to meet in phase in the middle of the interconnecting cable, resulting in an electrical short circuit at the loops. (b) A push–pull signal will cause voltages to meet out of phase in the middle of the interconnecting cable, resulting in an electrical open circuit at the loops.

to all four loops. However, only one set of loops should be excited at a time, and this is accomplished in the following way:

To each of the four loops is connected a transmission line $\lambda/4$ long (switching traps) and is terminated in a switching diode. When a pair of diodes are open (i.e., not conducting), we obtain short circuits at these two loop inputs (i.e., the energy cannot be radiated by these particular loops). Contrarily, if a pair of diodes act as short circuit (i.e., conducting), the impedance of the switching traps at the loop inputs is infinite and will permit energy transfer between the loops to the feeding lines. Note an additional feature in Figure B.11: that the transmission lines from the loops to the two branch points B and B' are all about $\lambda/4$ long. The reason for this is simply that when the switching traps act like short circuits at the

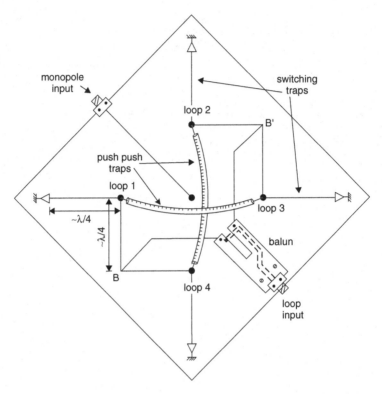

Figure B.11 Bottom view of the actual layout of the circuit board feeding the monopole and the four loops from the balun. Also note the switching traps activating the loops. In the model, screws were used instead of diodes.

input of one set of the loops, a high impedance will be produced at point B (i.e., the energy transfer to the other set of loops will not be disturbed).

B.4.3 Phase Reversal in the Balun

As can be seen from the description above, the switching traps determine which set of loops is going to radiate. If we further excite the monopole with proper amplitude and phase, a beam will be formed in one of the four directions east, west, north, or south. To shift the beam from, for example, north to south, we merely have to change the phase of loops 2 and 4 180° with respect to the monopole. This could be accomplished by a simple λ/2 transmission line, which could be short-circuited by a diode. However, due to the relatively large bandwidth desired (ca. 25%), the electrical length of the transmission line would probably vary more over the frequency band than could be tolerated in order to maintain a good null in the back direction of the pattern (see later). This problem is avoided by performing the phase switching in the balun as illustrated in Figure B.12. Instead of a single inner conductor we have split it up into two in parallel, each of which has a total length of $\frac{3}{4}\lambda$. We have further connected a pair of transmission lines $\lambda/4$ long at points B and B′, and each terminated in a switching diode. Let us now assume that the diode to the left is open-circuited and the one to the right is short-circuited. That will result in a short circuit across the transmission line at point B, while the energy can pass unobstructed along the feedline going through point B′. Furthermore, since the distance from point B to the branch point A is about $\lambda/4$, the short circuit at B will produce a high impedance at point A (i.e., the energy will flow freely through the transmission line to the right). Finally, since the distance from point B to point C (at the balanced terminal at the top) is about $\lambda/2$, the short circuit at B will also produce a short circuit at point C between the inner and outer conductor, which is precisely what is required for proper operation of a Roberts balun

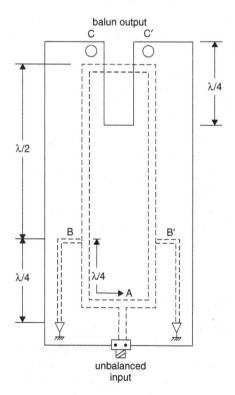

Figure B.12 Modified Roberts balun provided with switch traps, enabling us to change the phase 180°.

(in fact, the frequency dependence of the $\sim\lambda/2$ transmission line will partly compensate for the frequency dependence of the $\sim\lambda/4$ fork at the balanced output).

By reversing the conducting condition of the two diodes in Figure B.12, we can lead the power transfer along the transmission to the left. The important point now is to note that this will change the phase of the signal at the balanced output by exactly $180°$ (i.e., the beam will rotate by $180°$ in the horizontal plane). Note further that due to complete symmetry of the balun in Figure B.12, the frequency sensitivity of the balanced output is the same for the two phase positions. As we shall see later, this is important in order to maintain a good null in the back direction of the radiation pattern.

B.4.4 Final Execution of Design 2b

In Figure B.13 we show the final version of our preferred design 2b. We notice the top-loaded monopole in the middle of the cavity. It is fed from the upper N-connector at the bottom to the left via a 100-Ω microstrip line in series with a capacitor. The purpose of the latter is to invert the monopole impedance while the 100-Ω microstrip line transforms the impedance very nicely into about 50 Ω (for details about this technique, see Appendix B in ref. 2). Further, the four loops are fed from the lower N-connector at the bottom to the left (only two loops are excited at a time). We clearly see the Roberts balun as well as one of the push–push traps.

Finally, we feed the two N-connectors from a specially designed hybrid. By measuring the fields radiated separately from the monopole and from a pair of loops over the entire frequency band, it was determined that the best fit would be obtained by using a 4- to 2-dB hybrid. This little jewel was not only designed but also built by use of an Exacto knife in a single night by Clayton Larson, one of the author's former students. In fact, this antenna would never had been realized without Clayton at my side. We had fun!

B.4.5 Radiation Pattern

Typical radiation pattern in the horizontal plane is shown in Figure B.14 for the midrange frequencies. The complete set of patterns can be found in the original report [1]. However, we do show in Figure B.15 the front-to-back ratio over the entire band 1.0 to 1.25 GHz.

Finally, shown in Figure B.16 are the vertical radiation patterns in the midfrequency range. A complete set is given in ref. 1.

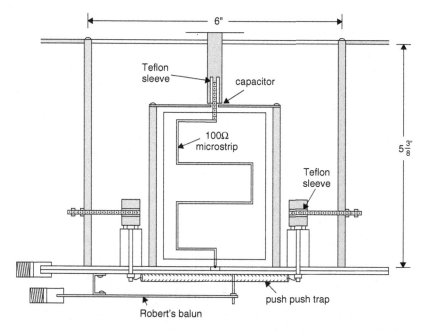

Figure B.13   Final model of design 2b.

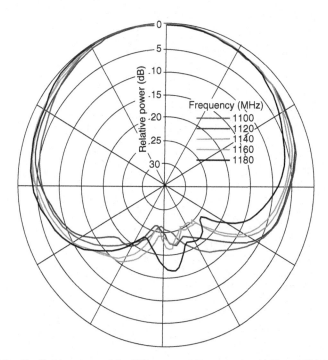

Figure B.14   Cardioid pattern of the IFF antenna. Frequencies 1100 to 1190 MHz. Taken at elevation about $8°$ above horizon. Loop $1+3$ activated. Vertical polarization.

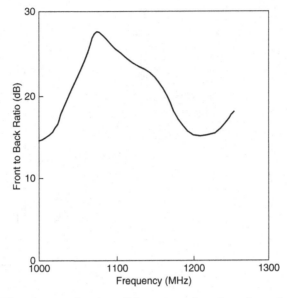

Figure B.15 FB ratio as a function of frequency taken from the pattern measured in Figure B.14.

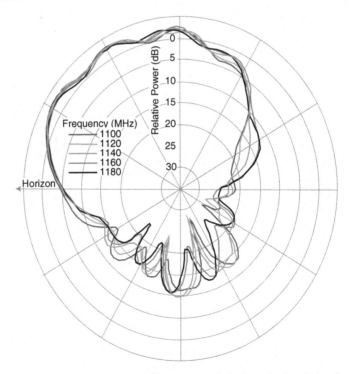

Figure B.16 Vertical pattern of the IFF antenna taken through the plane of symmetry. Frequencies 1100 to 1180 MHz. Loop 1 + 3 activated. Vertical polarization.

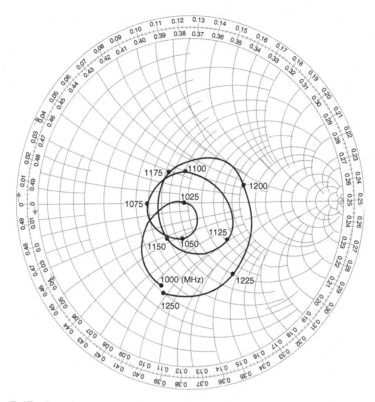

Figure B.17 Impedance seen at the 4-dB hybrid connected to the monopole and a pair of loops via the balun.

B.4.6   Impedance

The impedance as seen at the input of the 4-dB hybrid is shown in Figure B.17. It is lacking slightly at the first 0.01 GHz as well as the last 0.05 GHz. However, these frequencies were actually outside the frequency band designed.

B.5   CONCLUSIONS

We produced an antenna comprised of four loops placed in a circular cavity with a top-loaded monopole placed in the middle. Such an antenna could easily produce a broad steerable beam with vertical polarization in the frequency range 1.0 to 1.25 GHz. The radiation in the back sector was 15 to 25 dB below the main beam, depending on the frequency (20 dB or more between 1030 and 1150 MHz). The measured gain was about 2 dB higher than that of a standard IFF antenna. The VSWR was less than 2 : 1 over 1010 to 1200 MHz. The complete antenna was about 15 cm deep. However, a total depth of about 5 cm seems possible, as discussed in Appendix B of ref. 1. The push–push traps was covered by a patent [3].

REFERENCES

[1] B. A. Munk and C. J. Larson, A cavity-type broadband antenna with a steerable cardioid pattern, *Final Rep. 711559-2, ElectroScience Laboratory*, Ohio State University, Dec. 1979.

[2] B. A. Munk, *Finite Arrays and FSS*, Wiley, Hoboken, NJ, 2003.

[3] B. A. Munk and C. J. Larson, Broadband multi-element antenna, U.S. patent 4,540,988, Sept. 10 1985.

APPENDIX C
How to Measure the Characteristic Impedance and Attenuation of a Cable

C.1 BACKGROUND

Several of the antenna designs presented in this book and elsewhere have used matching networks put together with cable sections of various characteristic impedance. Designing such sections, whether coaxial, twin lead, or microstrip, is well known and will therefore not be discussed here.

In the case of coaxial cable sections, the practical execution may be in the form of a rigid outer conductor into which we insert a suitable inner conductor supported by either beads or solid dielectric (foam is often used). Another approach is simply to draw the inner conductor out of a suitable coaxial cable and, instead, insert a suitable thinner inner conductor leading to a higher characteristic impedance. Although this sounds somewhat crude and primitive, I have watched technicians perform such an "operation" several times with very good success.

Quite often, it is desirable to verify your calculations of the characteristic impedance by measurements. Although many practitioners will be quite familiar with this problem, I am quite often asked to comment on this topic. Thus, since this definitely is important when producing antennas as alternatives to those made of exotic materials that very often cannot be produced, I thought it appropriate to give the following brief discussion.

Measurements of the characteristic impedance typically start with the input impedance of a cable section terminated in some load impedance. More specifically, we show in the insert of Figure C.1 a transmission line of length l, propagation constant β and characteristic impedance Z_0. It is terminated in a load impedance Z_L, resulting in the reflection coefficient

Metamaterials: Critique and Alternatives, By Ben A. Munk
Copyright © 2009 John Wiley & Sons, Inc.

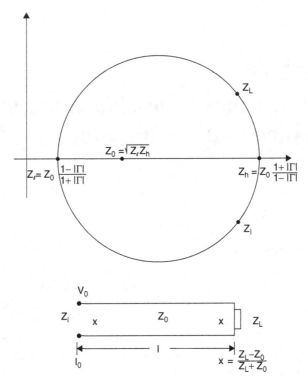

Figure C.1 The input impedance Z_i moves on a circle determined by Z_l and Z_h as indicated in the figure. The characteristic impedance is determined by $Z_0 = \sqrt{Z_l Z_h}$.

$$\Gamma = \frac{Z_L - Z_0}{Z_L + Z_0} \qquad (C.1)$$

The expression for the input impedance Z_i has many forms. However, the author's favored form is readily obtained by noting that when the voltage V_0 is applied to the input terminals, a signal will travel toward the load Z_L, where it will be reflected as ΓV_0. When it arrives back at the input terminals, the total phase delay will be $e^{-j2\beta l}$; that is, the total voltage V_{tot} at the input terminal is

$$V_{\text{tot}} = V_0 \left(1 + \Gamma e^{-j2\beta l}\right) \qquad (C.2)$$

Similarly, the current I_0 will be reflected as $-\Gamma I_0$, leading to a total current I_{tot} at the input terminals

$$I_{\text{tot}} = I_0 \left(1 - \Gamma e^{-j2\beta l}\right) \qquad (C.3)$$

Thus, the input impedance is

$$Z_i = \frac{V_{tot}}{I_{tot}} = \frac{V_0}{I_0}\frac{1 + \Gamma e^{-j2\beta l}}{1 - \Gamma e^{-j2\beta l}} \quad \text{or} \quad Z_i = Z_0\frac{1 + |\Gamma|e^{j\varphi_r}e^{-j2\beta l}}{1 - |\Gamma|e^{j\varphi_r}e^{-j2\beta l}} \quad \text{(C.4)}$$

where $Z_0 = V_0/I_0$.

The beauty of (C.4) is that it constitutes a bilinear relationship between the independent variable $z = e^{-j2\beta l}$ and the dependent variable Z_i. Thus, as is well known [1], when z moves along a circle (which it obviously does as $2\beta l$ varies with either frequency or cable length), the input impedance Z_i will also be located on a circle, which is easy to see by inspection of (C.4). Quite simply, when

$$e^{j\varphi_r}e^{-j2\beta l} = 1 \quad \text{(C.5)}$$

the largest numerical (and real) value of Z_i becomes (see Figure C.1)

$$Z_h = Z_0\frac{1 + |\Gamma|}{1 - |\Gamma|} \quad \text{(C.6)}$$

Similarly, for

$$e^{j\varphi_r}e^{-j2\beta l} = -1 \quad \text{(C.7)}$$

the smallest (and real) value of Z_i is

$$Z_l = Z_0\frac{1 - |\Gamma|}{1 + |\Gamma|}. \quad \text{(C.8)}$$

From equations (C.6) and (C.8),

$$Z_h Z_l = Z_0^2 \quad \text{or} \quad Z_0 = \sqrt{Z_h Z_l}. \quad \text{(C.9)}$$

Thus, to determine Z_0 we merely plot the input impedance Z_i as a function of frequency in the complex plane when the cable is terminated in an arbitrary impedance Z_L. Z_i should follow a circle that intersects the real axis in Z_h and Z_l. The characteristic impedance Z_0 is then given by (C.9).

Note: It is not necessary to know the load impedance, Z_L. But if you do, the circle for the input impedance must go through Z_L eventually compensated for cable loss (see below).

Note: It was implied above that Z_L could be arbitrary as long as it does not change with frequency. Thus, we conclude immediately that Z_L should be real. Also, values of Z_L close to Z_0 usually provide us with better accuracy. In fact, using a short circuit and an open circuit usually does not work satisfactorily. (Open circuits cannot, in general, be trusted. They radiate.)

C.2 INPUT CONNECTOR EFFECT

Quite often it turns out that the input circle has its center located somewhat off the real axis. An example of such a case is shown in Figure C.2. There can be various reasons for this displacement, but quite often it is caused by a mismatch between the connector and the cable. Typically, the connector has a characteristic impedance of 50 Ω, while the cable has a higher characteristic impedance. This will cause the two impedances Z_l and Z_h to be relocated to positions Z'_l and Z'_h, respectively. For the sake of simplicity, let us assume that the distances $OZ_l = OZ'_l$ and $OZ_h = OZ'_h$; that is, from (C.9) we conclude that

$$Z_0 = \sqrt{OZ'_l \cdot OZ'_h} \tag{C.10}$$

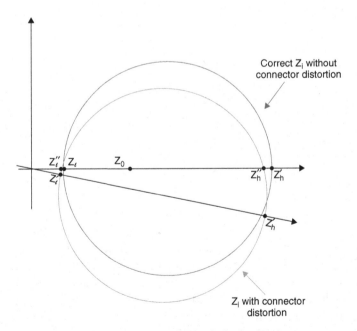

Figure C.2 Typical displacement of the input circle for Z_i due to a connector with characteristic impedance different from Z_0. The characteristic impedance is given approximately by $Z_0 = \sqrt{Z''_l Z''_h}$.

Let us further denote the points where the dislocated impedance circle crosses the real axis by Z''_l and Z''_h. Then from the theorem "power of the circle" [1] we also have

$$OZ''_l \cdot OZ''_h = OZ'_l \cdot OZ'_h \tag{C.11}$$

And finally, from (C.10) and (C.11),

$$Z_0 \sim \sqrt{OZ''_l \cdot OZ''_h} \tag{C.12}$$

In other words, if the dislocated circle crosses the real axis in the impedances Z''_l and Z''_h, the characteristic impedance Z_0 is given *approximately* by

$$Z_0 \sim \sqrt{Z''_l Z''_h} \tag{C.13}$$

C.3   DO THE FORMULAS HOLD IN THE SMITH CHARTS?

So far we have considered the usual complex plane with a rectangular coordinate system. In such a system we have made use of the fact that the impedance Z_l is proportional to the distance OZ_l and the analog for Z_h. This is not the case when we work in the Smith chart, and the question arises: Will the formulas above work in the Smith chart? This situation is illustrated in Figure C.3. Actually, all we have to do is merely read the numbers for Z_l and Z_h directly in the Smith chart. Substituting these values in (C.10) will yield the correct value for Z_0, and in the case of (C.13), an approximate value.

Note: How the Smith chart is normalized with respect to Z_1 is immaterial as long as we read the actual values for Z_l and Z_h.

C.4   HOW TO MEASURE THE CABLE LOSS

Note: In contrast to the measurements performed above, the determination of the cable loss can only be determined in a Smith chart normalized to the characteristic impedance Z_0 of the cable in question.

If we terminate a cable of length l in a short, the input impedance for the lossless case is obtained by rotating $2\beta l$ along the rim of the Smith

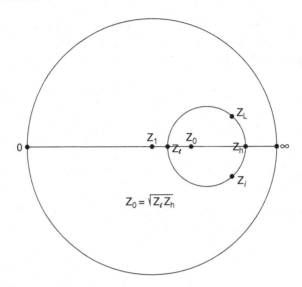

Figure C.3 Typical input impedance circle when located in a Smith chart with reference to $Z_1 \neq Z_0$.

chart as illustrated in Figure C.4. If the cable has an attenuation of α neper/meter, the distance from the center of the Smith chart is reduced to

$$b = ae^{-2\alpha l} \qquad (C.14)$$

where a is the radium of the Smith chart as illustrated in Figure C.4. From (C.14) we readily see that the total two-way attenuation for a cable of length l is

$$2\alpha l(\text{dB}) = 20 \, \log a/b(\text{dB})$$

where α is now measured in dB/m. Thus, the one-way attenuation for the cable is

$$\alpha(\text{dB/m}) = \frac{1}{l(\text{m})} 10 \, \log a/b(\text{dB/m}) \qquad (C.15)$$

Thus, the attenuation of a cable of length l and characteristic impedance Z_0 is obtained by plotting the input impedance Z_i in a Smith chart normalized to Z_0. This impedance will be located on a circle with its center at the center of the Smith chart. By measuring the ratio between this impedance circle and the radius of the Smith chart, we can immediately obtain the attenuation α (dB/m) from (C.15).

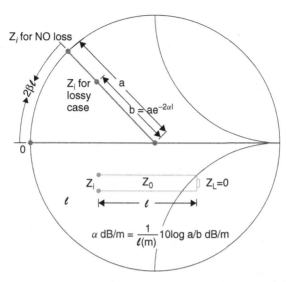

Figure C.4 To determine the attenuation of a cable of length $l(m)$, we plot the input impedance Z_i of the short-circuited cable in a Smith chart normalized to Z_0. The attenuation is obtained from $\alpha(\mathrm{dB/m}) = [1/l(\mathrm{m})]10\,\log(\frac{a}{b})(\mathrm{dB/m})$.

REFERENCE

[1] B. A. Munk, *Frequency Selective Surfaces: Theory and Design*, Wiley, New York, 2000, App. A.

APPENDIX D
Can Negative Refraction Be Observed Using a Wedge of Lossy Material?

D.1 INTRODUCTION

Throughout this book we have steadfastly claimed that a negative index of refraction does not seem possible. However, a recent paper by Sanz et al. [1] showed that a glass wedge with loss but without negative parameters could produce negative refraction at 320 nm. Garcia and Nieto-Vesperinas [2] showed analytically that below negative index media (NIM) lattice, cutoff ε is imaginary and losses dominate. These papers inspired R. C. Hansen to investigate the possibility of a negative index of refraction at microwave frequencies [3]. He exposed a dielectric wedge of very lossy material to an incident field produced by a linear array of isotropic line sources parallel to the wedge edge and spaced $\lambda/4$ to avoid grating lobe complications. Calculation of the field transmitted seems to indicate that a negative index of refraction is possible.

The purpose of this appendix is to show that significant fields in the negative sector can indeed be obtained using a wedge of very lossy material. However, as we show, it is not caused by negative refraction but probably by a broadening of the beam transmitted. We start our investigation by reviewing certain fundamentals.

D.2 REFRACTION FOR PLANAR SLABS

In this section the transmission of an incident plane wave through a planar dielectric slab located in air is considered. We denote the angle of incidence by θ_{in}, the angle of refraction by θ_{out}, and the angle inside the

Metamaterials: Critique and Alternatives, By Ben A. Munk
Copyright © 2009 John Wiley & Sons, Inc.

dielectric by θ_{die}, as shown in Figure D.1. θ_{in} and θ_{die} are related by Snell's law:

$$\frac{\sin \theta_{die}}{\sin \theta_{in}} = \frac{1}{\sqrt{\mu \varepsilon}} \tag{D.1}$$

The lossy case is somewhat more complicated by the fact that we no longer are dealing with simple plane waves inside the slab, as shown in Figure D.2. Here the planes of equal phase are still orthogonal to the direction of propagation, θ_{die}, while the planes of equal amplitude are parallel with the input face. However, θ_{die} is no longer obtained from the simple Snell's law (D.1) but from Stratton [4]:

$$\frac{\sin \theta_{die}}{\sin \theta_{in}} = \frac{\beta_{air}}{\sqrt{q^2 + \beta_{air}^2 \sin^2 \theta_{in}}} \tag{D.2}$$

where β_{air} is the propagation constant in air,

$$q = \rho(\beta_{die} \cos \gamma - \alpha_{die} \sin \gamma), \tag{D.3}$$

and α_{die} is the attenuation (N/m) in the dielectric. Further, ρ and γ are defined by

$$\cos \theta_1 = \sqrt{1 - (a^2 - b^2 + j2ab) \sin^2 \theta_{in}} = \rho e^{-j\gamma} \tag{D.4}$$

where θ_1 is the complex angle of the refracted field in the dielectric obtained from Snell's law by matching the phase velocities. Further,

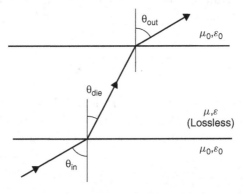

Figure D.1   Definition of the incidence angle θ_{in}, the refraction angle θ_{out}, and the angle θ_{die} inside a planar lossless dielectric slab.

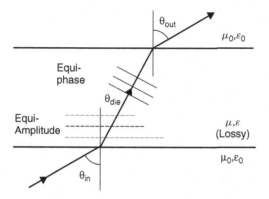

Figure D.2 Inside a lossy dielectric slab the field consists of inhomogeneous plane waves with the equiphase planes orthogonal to the direction θ_{die}, whereas the equiamplitude planes are parallel to the input plane. Note: $\theta_{\text{in}} \equiv \theta_{\text{out}}$ regardless of loss.

denoting the propagation constant in dielectric by β_{die}, we have

$$a = \frac{\beta_{\text{air}}\beta_{\text{die}}}{\alpha_{\text{die}}^2 + \beta_{\text{die}}^2} \tag{D.5}$$

$$b = \frac{\beta_{\text{air}}\alpha_{\text{die}}}{\alpha_{\text{die}}^2 + \beta_{\text{die}}^2} \tag{D.6}$$

For details, see Section 9.8 in ref. 4.

Further, from equations (48) and (49) in Section 5.2 in ref. 4, we have, for a lossy dielectric,

$$\alpha_{\text{die}} = \omega \left[\frac{\mu\varepsilon}{2} \left(\sqrt{1 + \frac{\sigma^2}{\varepsilon^2\omega^2}} - 1 \right) \right]^{1/2} \tag{D.7}$$

$$\beta_{\text{die}} = \omega \left[\frac{\mu\varepsilon}{2} \left(\sqrt{1 + \frac{\sigma^2}{\varepsilon^2\omega^2}} + 1 \right) \right]^{1/2} \tag{D.8}$$

where σ is the conductivity of the dielectric. From equations (D.7) and (D.8),

$$\alpha_{\text{die}}^2 + \beta_{\text{die}}^2 = \omega^2\mu\varepsilon\sqrt{1 + \frac{\sigma^2}{\varepsilon^2\omega^2}} \tag{D.9}$$

Substituting equation (D.9) into (D.5) and (D.6) yields

$$a = \frac{\beta_{air}\beta_{die}}{\omega^2\mu\varepsilon\sqrt{1+\sigma^2/\varepsilon^2\omega^2}} \tag{D.10}$$

$$b = \frac{\beta_{air}\alpha_{die}}{\omega^2\mu\varepsilon\sqrt{1+\sigma^2/\varepsilon^2\omega^2}} \tag{D.11}$$

Further, from equations (D.10) and (D.11) and by application of (D.7) and (D.8), we have

$$a^2 - b^2 = \frac{\beta_{air}^2}{(\omega^2\mu\varepsilon\sqrt{1+\sigma^2/\varepsilon^2\omega^2})^2}(\beta_{die}^2 - \alpha_{die}^2)$$

$$= \frac{\beta_{air}^2}{\omega^2\mu\varepsilon\left(1+\sigma^2/\varepsilon^2\omega^2\right)} \tag{D.12}$$

and

$$2ab = 2\frac{\beta_{air}^2\beta_{die}\alpha_{die}}{\left[\omega^2\mu\varepsilon\sqrt{1+\sigma^2/\varepsilon^2\omega^2}\right]^2}$$

$$= 2\beta_{air}^2\frac{\omega^2\left[(\mu\varepsilon/2)\left(\sqrt{1+\sigma^2/\varepsilon^2\omega^2}+1\right)\right]^{1/2}\left[\frac{\mu\varepsilon}{2}\left(\sqrt{1+\sigma^2/\varepsilon^2\omega^2}-1\right)\right]^{1/2}}{\left[\omega^2\mu\varepsilon\sqrt{1+\sigma^2/\varepsilon^2\omega^2}\right]^2}$$

$$= \frac{\beta_{air}^2(\sigma/\varepsilon\omega)}{\omega^2\mu\varepsilon\left(1+\sigma^2/\varepsilon^2\omega^2\right)} \ll 1 \qquad \text{for} \frac{\sigma}{\varepsilon\omega} \gg 1 \tag{D.13}$$

Further applying (D.12) in (D.13) yields

$$\frac{2ab}{a^2-b^2} = \frac{\sigma}{\varepsilon\omega}$$

That is, for very lossy dielectric where $\sigma/\varepsilon\omega \gg 1$, we have

$$2ab \gg a^2 - b^2 \tag{D.14}$$

Applying (D.14) to (D.4) yields

$$\rho\varepsilon^{-j\gamma} = \sqrt{1 - (a^2 - b^2 + jab)\sin^2\theta_{in}}$$

$$\simeq \sqrt{j - j2ab\sin^2\theta_{in}} \tag{D.15}$$

and since $2ab \ll 1$ according to (D.13),

$$\rho \varepsilon^{-j\gamma} \sim 1 - jab \sin^2 \theta \qquad (D.16)$$

yielding $\rho \sim 1$ and $\gamma \sim 0$. Substituting these values into (D.3) yields

$$q = \rho(\beta_{\text{die}} \cos \gamma - \alpha_{\text{die}} \sin \gamma) \sim \beta_{\text{die}} \qquad (D.17)$$

Finally, substituting (D.8) into (D.17) yields

$$q \sim \sqrt{\frac{\omega \mu \sigma}{2}} \qquad (D.18)$$

which agrees with (64) on p. 504 in ref. 4. Finally, substituting (D.18) into (D.2) yields

$$\frac{\sin \theta_{\text{die}}}{\sin \theta_{\epsilon}} = \frac{\beta_{\text{air}}}{\sqrt{\omega \mu \sigma / 2}} \qquad (D.19)$$

We see from (D.19) that the angle of refraction θ_{die} inside the dielectric is small, indeed, but *never* negative. Frankly, since nobody, to the best of this author's knowledge, has ever suggested that a negative index of refraction exists for μ, $\varepsilon > 0$, it may seem that the rather excessive proof above was a waste of time. Perhaps! Nevertheless, here it is, in case someone doubts it.

D.3   WEDGE-SHAPED DIELECTRIC

Instead of planar dielectric slabs, let us next consider the wedge-shaped cases. The lossless case is shown in Figure D.3, where we have assumed that θ_{die} is at least approximately equal to the planar lossless case considered earlier. Inspection of the figure readily shows that the new θ_{out} for the wedge-shaped case is somewhat larger than for the planar case. This is supported further by the fact that the rays closest to the wedge edge go through less dielectric and are therefore less delayed, leading to further tilt (i.e., a larger value of θ_{out}). However, if the direction is measured with respect to the input plane, the wedge case and the planar case can be larger or smaller than the planar case depending on the wedge angle, Δ, but never negative.

In the lossy case depicted in Figure D.4, two important points should be mentioned. First, the direction of the rays inside the wedge will be closely

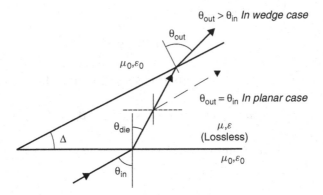

Figure D.3 In the lossless wedge case, θ_{in} and θ_{die} are defined as in the planar case, whereas θ_{out} is measured from the normal to the output face (i.e., $\theta_{out} > \theta_{in}$). However, if θ_{out} is measured from the normal to the input plane, we can have $\theta_{out} > <\theta_{in}$ but never $\theta_{out} < 0$.

aligned with the normal to the input plane. Again, as in the lossless case, this leads to a slightly greater value of θ_{out} for the wedge case. However, if we measure θ_{out} from the normal to the input plane, θ_{out} can be larger or smaller than θ_{in} but never negative. The second point is that Hansen performed his experiment not with an incident plane wave but rather, a beam produced by a finite array with an aperture of about 4λ, producing a beamwidth of about $15°$. This will result in a finite aperture illumination at the output surface of the wedge. The field radiated into the upper space

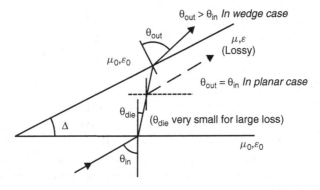

Figure D.4 In the lossy wedge case, θ_{in}, θ_{die}, and θ_{out} are basically defined as in the lossless case, with the same restrictions except that θ_{die} is very small but never negative. Also, θ_{die} is different from the lossless case.

can be obtained from the equivalence theorem by calculating the field radiated from this aperture.

In the lossless case this aperture field would be a fairly accurate replica of the array beam and no particular unexpected feature would be observed by evaluation of the radiation pattern. In the lossy case, however, it is important to realize that the part of the aperture to the left has a much stronger field than the part to the right because the longer path in the lossy wedge attenuates more. We do not have all the information necessary to calculate the exact aperture distribution in the lossy case. However, we shall make an approximate argument that very clearly indicates the effect of having a highly asymmetric aperture illumination rather than a symmetric one.

D.4 ASYMMETRIC APERTURE DISTRIBUTIONS IN GENERAL

A simple asymmetric aperture distribution is the triangular shape shown in Figure D.5a. It can, like all arbitrary distributions, be decomposed into an even and an odd distribution. In this case there will be the rectangular shape shown in Figure D.5b and the double triangular shape shown in Figure D.5c. The radiation patterns for these two components are shown in Figure D.6a and b, respectively. The symmetric yields the well-known $(\sin x)/x$ pattern, where the sidelobe level starts at around 13 dB below the main beam. The double triangular shape yields two beams located on each side of the main beam and about 7 dB below it. Note that the odd component is always in quadrature with the even component. Thus, the sum of these two components yielding the total field, as shown in Figure D.6c can never have a null.

Due to the high loss in the wedge, it is entirely possible that the actual aperture field is located strongly to the left as indicated in Figure D.7. In that case, the amplitude of the radiation pattern will basically be twice as wide as for the aperture shown in Figure D.6c. Thus, we could obtain readily a strong signal in the "negative" sector and misinterpret this as negative refraction.

D.5 CONCLUSIONS

Although the discussion above has, at times, been somewhat heuristic, for reasons explained in the footnote, we are nevertheless able to draw some very important conclusions about refraction from a very lossy wedge.

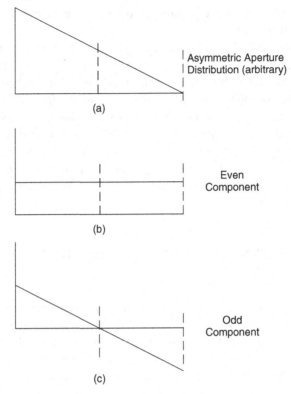

Figure D.5 Any arbitrary aperture distribution, as in this case a triangular distribution as seen in part (a) can always be decomposed into an even distribution as shown in part (b) and an odd distribution as shown in part (c).

When the input angle θ_{in} is small, the output angle θ_{out} measured with respect to the input face can be very small as well, depending on the wedge angle, Δ. However, it can *never* be negative, neither in the lossless nor the lossy case (see Figures D.3 and D.4).

When the wedge is exposed to an antenna beam rather than an incident plane wave, the field transmitted through the wedge can, by the equivalence theorem, be obtained from an equivalent aperture on the top surface of the wedge. In the lossless case the field transmitted will be a good replica of the incident beam. In the lossy case, however, the equivalent aperture field will be highly asymmetric, due to the larger attenuation when going through the thicker part of the wedge.

This new asymmetric aperture field was decomposed into an even and an odd component. Since these two components are always in quadrature, their sum, i.e., the total pattern, can never have a null. If we assume further

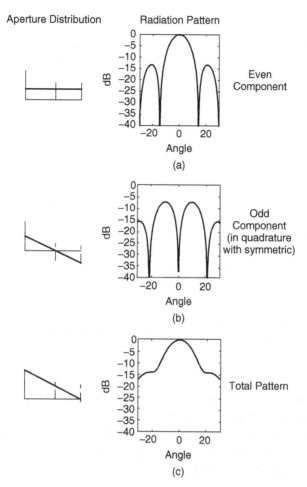

Figure D.6 (a) The radiation pattern $(\sin x)/x$ of the even aperture distribution. The sidelobe level is down at least 13 dB below the main beam. (b) The radiation pattern for odd distribution, in this case double triangular shape. Produced two beams about 7 dB below the main beam and with the same beamwidth, one 90° ahead, the other 90° behind the main beam. (c) The total pattern is the sum of parts (a) and (b). It's always higher than the highest component. The calculation of these curves by Jens Munk is gratefully acknowledged.

that the asymmetric aperture distribution fills only half the original lossless aperture, we obtain a new beamwidth that is roughly twice as wide as in the lossless case. Thus, part of the beam can readily show up in the "negative" sector, and be interpreted as negative refraction.

These findings substantiate to a high degree Hansen's calculated result: The angle of incidence θ_i should be small, as shown in Figure 3 of

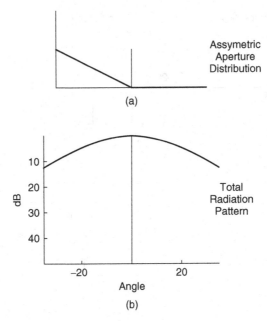

(a)

(b)

Figure D.7 A realistic aperture distribution for very loss wedge. Beamwidth is twice the value shown in Figure D.6c.

Hansen's paper [3], the wedge material very lossy, and the incident field an antenna beam rather than a plane wave. Obviously, I agree wholeheartedly with Hansen. On one point I must respectfully disagree, however. This phenomenon should not be referred to as negative refraction—simply because it is not. It is positive.

REFERENCES

[1] M. Sanz et al. , Transmission measurements in wedge-shaped absorbing samples: an experiment for observing negative refraction, *Phys. Rev.*, vol. 67, pp. 1–4, 2003.

[2] N. Garcia and M. Nieto-Vesperinas, Is there an experimental verification of a negative index of refraction yet?, *Opt. Lett.*, vol. 27, pp. 885–887, June 2002.

[3] R. C. Hansen, Negative refraction without negative index, *IEEE Trans. Antennas Propag.*, vol. 56, Feb. 2008.

[4] J. A. Stratton, *Electromagnetic Theory*, McGraw-Hill, New York, 1941, pp. 490–505.

INDEX

Metamaterials: Critique and Alternatives, By Ben A. Munk
Copyright © 2009 John Wiley & Sons, Inc.

Printed in the United States
By Bookmasters